T0076637

Two from One

Two from One

A Short Introduction to Cell Division Mechanisms

Michael Polymenis

Department of Biochemistry and Biophysics
Texas A&M University
College Station, Texas

Illustrations by Athená Polymenis

WILEY

This edition first published 2023
© 2023 John Wiley & Sons Ltd

All rights reserved. No part of this publication may be reproduced, stored in a retrieval system, or transmitted, in any form or by any means, electronic, mechanical, photocopying, recording or otherwise, except as permitted by law. Advice on how to obtain permission to reuse material from this title is available at http://www.wiley.com/go/permissions.

The right of Michael Polymenis to be identified as the author of this work has been asserted in accordance with law.

Registered Offices
John Wiley & Sons, Inc., 111 River Street, Hoboken, NJ 07030, USA
John Wiley & Sons Ltd, The Atrium, Southern Gate, Chichester, West Sussex, PO19 8SQ, UK

Editorial Office
9600 Garsington Road, Oxford, OX4 2DQ, UK

For details of our global editorial offices, customer services, and more information about Wiley products visit us at www.wiley.com.

Wiley also publishes its books in a variety of electronic formats and by print-on-demand. Some content that appears in standard print versions of this book may not be available in other formats.

Trademarks: Wiley and the Wiley logo are trademarks or registered trademarks of John Wiley & Sons, Inc. and/or its affiliates in the United States and other countries and may not be used without written permission. All other trademarks are the property of their respective owners. John Wiley & Sons, Inc. is not associated with any product or vendor mentioned in this book.

Limit of Liability/Disclaimer of Warranty
In view of ongoing research, equipment modifications, changes in governmental regulations, and the constant flow of information relating to the use of experimental reagents, equipment, and devices, the reader is urged to review and evaluate the information provided in the package insert or instructions for each chemical, piece of equipment, reagent, or device for, among other things, any changes in the instructions or indication of usage and for added warnings and precautions. While the publisher and authors have used their best efforts in preparing this work, they make no representations or warranties with respect to the accuracy or completeness of the contents of this work and specifically disclaim all warranties, including without limitation any implied warranties of merchantability or fitness for a particular purpose. No warranty may be created or extended by sales representatives, written sales materials or promotional statements for this work. The fact that an organization, website, or product is referred to in this work as a citation and/or potential source of further information does not mean that the publisher and authors endorse the information or services the organization, website, or product may provide or recommendations it may make. This work is sold with the understanding that the publisher is not engaged in rendering professional services. The advice and strategies contained herein may not be suitable for your situation. You should consult with a specialist where appropriate. Further, readers should be aware that websites listed in this work may have changed or disappeared between when this work was written and when it is read. Neither the publisher nor authors shall be liable for any loss of profit or any other commercial damages, including but not limited to special, incidental, consequential, or other damages.

Library of Congress Cataloging-in-Publication Data
Names: Polymenis, Michael, author.
Title: Two from one : a short introduction to cell division mechanisms /
    Michael Polymenis, Professor, Department of Biochemistry and Biophysics,
    Texas A&M University, College Station, Texas.
Description: First edition. | Hoboken, NJ, USA : John Wiley & Sons, Inc.,
    2023. | Includes bibliographical references and index.
Identifiers: LCCN 2022038433 (print) | LCCN 2022038434 (ebook) | ISBN
    9781119930143 (Paperback) | ISBN 9781119930150 (Adobe pdf) | ISBN
    9781119930167 (Epub)
Subjects: LCSH: Cell division. | Cell cycle.
Classification: LCC QH605 .P57 2023 (print) | LCC QH605 (ebook) | DDC
    571.8/44–dc23/eng/20220920
LC record available at https://lccn.loc.gov/2022038433
LC ebook record available at https://lccn.loc.gov/2022038434

Cover Design: Wiley
Cover Image: Reproduced under CC BY license, Figure 3 © Siegel D, Kepa JK, Ross D (2012)

Set in 11.5/13.5pt STIXTwoText by Straive, Pondicherry, India

For Helene, my best friend, lover, wife, and mother of our children, who is dealing with the catastrophic consequences of too much cell division.

Contents

Foreword

This book is about cell division, the basis of all life on this planet. It is based on the material covered in a short, five-week course developed by the author. It was designed for first-year graduate students in the life sciences or undergraduate juniors and seniors, who have some general biology and biochemistry background, but not much beyond that. The book is meant to be a solid step in learning about the subject, not the last for anyone interested in cell division. Learning about cell division is an excellent way to illustrate the unity and power of life sciences. The cell division field is a spectacular "melting pot," where general concepts and discoveries are synthesized from various systems and approaches.

I do not consider myself an expert in all things cell cycle. Although my research is in the general area, it is focused on a tiny slice of that area in the coupling of protein synthesis with cell division in yeast. Hence, I have probably overlooked some areas or overemphasized others. Some of these errors, hopefully not too many, may result from my poor knowledge of specific processes. The rest were mainly intentional, looking for ways to simplify the story without missing key pieces and telling it in a way that would make it more digestible to my students and other readers.

The text has been taught twice, revised, and simplified, based on student feedback, to be as accessible as possible. Emphasis is on general concepts. The "curse" of modern descriptions of cell division mechanisms is that they quickly morph into an "alphabet soup" of gene and protein names. There are fewer than a hundred such names on these pages. Hence, I must apologize for the many proteins and their cell cycle roles discovered by numerous scientists that are not covered here. I also wish to thank colleagues who pointed out errors of fact for the proteins and processes I do cover and apologize for any remaining mistakes. The book is not about providing the most comprehensive assembly of the current knowledge on cell division mechanisms. It can be read in a few hours by anyone with some interest in the topic and minimal undergraduate background. The book aims to outline the beauty and logic of the most vital biological process, making two cells from one.

My daughter, Athená Polymenis, "touched-up" most of the illustrations in the figures. Her talent for drawing is astonishing. I am glad she did not inherit my drawing skills.

I am very grateful to all the Department of Biochemistry and Biophysics students who commented on early versions of the text. They faced the material at the "frontlines," in the classroom. Their suggestions made a difference as the book evolved, especially toward simplifying it as much as possible. Likewise, the comments from anonymous reviewers were much appreciated. I thank all those at Wiley who transformed a preliminary draft into what you are reading. I am especially grateful to Julia Squarr, Senior Commissioning Editor, who saw potential in the concept and supported it. She kept me in line with her regular emails regarding decisions, deadlines, etc. I also thank Rosie Hayden, Joss Everett, and Naveen Kumaran Shanmugam for their help at various editorial steps. Finally, I thank my friends, lab members, and family who encouraged me to do this, even though it meant that, on occasion, they had to deal with an even more overworked and neurotic version of me.

—Michael Polymenis
Texas A&M University,
College Station, TX, USA

Preface

The cell cycle is what needs to happen from the time one cell is generated at the end of a cell division for that cell to divide into two. To make two cells from one, all the parts of one cell are duplicated and segregated into two cells. In the last 50 years, there have been several excellent monographs on cell division. The overwhelming emphasis in most of them is on the replication and segregation of the cell's chromosomes. Endowing the next generation of cells with the correct genetic material is the primary task a dividing cell must accomplish. But focusing only on chromosomes is like watching a movie with a single actor. After all, chromosomes make up only a small fraction of the cells' parts. As it turns out, when and how cells make and segregate their other components, even large molecules, such as proteins and RNAs, or smaller ones, such as lipids in membranes, is crucial in itself. It also impacts when and how cells duplicate and segregate their chromosomes.

But how does one go about describing all that information without getting lost in a blizzard of gene names and regulatory pathways? The book tries the following:

- A brief history of the cell cycle and its prominent landmarks, looking at cellular parts that are easily seen to duplicate and segregate.
- Discuss in some detail "bulk" cellular components that also need to be duplicated and segregated. But this process of cell "growth" is amorphous, with no apparent beginning or end. And yet, cell "growth" is coupled with cell division in ways that profoundly affect how fast cells multiply.

- Take a short detour to the current methods of monitoring cell division and their shortcomings. For example, some of these methods may perturb cellular physiology and the normal coordination between cell growth and division.
- Learn about the master control system of the cell cycle.
- Examine how cell cycle switches are put together and how they are turned on and off.
- Looking at gene expression patterns in the cell cycle, or at what goes up and down and when.
- After all the above, then we can better describe how the genome is duplicated and segregated.
- More than an afterthought, we will see how some organelles are duplicated and segregated.
- Finally, the last act, cytokinesis, yields two cells from one.

But other related topics will not be covered (e.g., meiosis, prokaryotic cell cycles, cell cycle controls during animal development, some unusual cell cycles). Not because these topics are uninteresting or unimportant. To keep the book short and more accessible, it can only focus on the common, universal aspects of eukaryotic cell division. Although a serious effort was made to minimize their number, some protein names are necessary. The final tally came to slightly less than a hundred. A table with all the protein names mentioned in the book and their function is given at the end.

Michael Polymenis, Texas A&M University,
College Station, TX, USA
January 2023

Symbols
and Abbreviations

[γ-^{32}P]-ATP	adenosine triphosphate, labeled on the gamma phosphate group with ^{32}P
⊣	inhibition
°C	degrees Celsius
Ʃ	time delay, hysteresis
1N	haploid genome content
2N	diploid genome content
^{3}H	tritium, a hydrogen atom that has two neutrons in the nucleus and one proton
APC	anaphase promoting complex
ATM	ataxia telangiectasia mutated
ATP	adenosine triphosphate
ATPγS	adenosine 5'-O-(3-thio)triphosphate
ATR	ataxia telangiectasia and Rad3 related
cdc	cell division cycle
Cdk	cyclin-dependent protein kinase
Cdk-as	analog-sensitive Cdk
CKI	Cdk inhibitor
dCTP	deoxycytidine triphosphate
DDK	Cdc7-Dbf4 kinase
DNA	deoxyribonucleic acid
dNTP	deoxyribonucleotide triphosphate
FACS	fluorescence-activated cell sorting
FDA	(United States) Food and Drug Administration
FUCCI	fluorescent ubiquitinated cell cycle indicator
G0	quiescent state, outside of the replicative cell cycle
G1 phase	the phase of the cell cycle lasting from the birth of the cell until initiation of DNA replication

G2 phase	the phase of the cell cycle when cells have fully replicated their DNA but have not started mitosis
GDP	guanosine diphosphate
GO	gene ontology
GTP	guanosine triphosphate
h	hours
I	input
k	specific proliferation (or volume growth) rate constant
K_M	Michaelis constant
M phase	mitosis
MPF	maturation promoting factor
mRNA	messenger RNA
MT	microtubule
MTOC	microtubule organizing center
N:C	nuclear:cytoplasmic ratio
nm	nanometer
nN	nanonewton
NPC	nuclear pore complex
O	output
PCNA	proliferating cell nuclear antigen
R^2	coefficient of determination
RNA	ribonucleic acid
RNase	ribonuclease
RO-3306	(5Z)-5-(quinolin-6-ylmethylidene)-2-(thiophen-2-ylmethylimino)-1,3--thiazolidin-4-one; a small molecule Cdk inhibitor
rRNA	ribosomal RNA
S phase	the phase of the cell cycle when cells replicate their DNA
SAC	spindle assembly checkpoint
SCF	Skp, Cullin, F-box containing
SPB	spindle pole body
T	total cell cycle time

T_{G1}	length of the G1 phase
TOR	target of rapamycin
tRNA	transfer RNA
Ub	ubiquitin

1 History and Context

1.1 From Cells to Their Nuclei

Until the last decades of the previous century, the history of the cell cycle is nearly the same as the history of the chromosome theory of heredity. And before chromosomes, we have to go back to when cells were first seen and then recognized as the unit of life. By the early seventeenth century, advances in lens grinding in the Netherlands allowed several individuals in Europe to experiment with and construct microscopes. Robert Hooke first used the word "cell" to describe what he saw when looking with his microscope at plant tissue. School children today have seen Hooke's famous drawings of cork sections. The drawings resemble a honeycomb, and they first appeared in Hooke's *Micrographia* in 1665. Hooke did not recognize that what he saw were the skeletal remains of all life's basic units. That realization did not happen until much later. In the decades after Hooke, numerous microscopists made a series of observations of cells. In 1719, looking at red cells from fish (which do not lose their nuclei as ours do), Antonie van Leeuwenhoek probably made the first drawings of the nucleus (1). The term "nucleus" was introduced in 1831 by Robert Brown (also of "Brownian" motion fame) when he used it to describe cells from orchids (1).

Two from One: A Short Introduction to Cell Division Mechanisms,
First Edition. Michael Polymenis.
© 2023 John Wiley & Sons Ltd. Published 2023 by John Wiley & Sons Ltd.

The discovery of the nucleus was highly significant. Not only because the nucleus contains the chromosomes (which had not been discovered at the time), but also because it is a visible cellular landmark. A marker that scientists can look for and monitor. Cellular landmarks, morphological or molecular, have driven and continue to drive cell cycle research. But just because a landmark is there does not necessarily mean that it may also make certain things happen. An example is the nucleolus and its role in the generation of new cells. Rudolf Wagner clearly described the nucleolus within the nucleus in 1835, which he assumed to be the "germinative spot" for cell formation. The nucleolus has important roles, but not in "seeding" new cells.

1.1.1 The Cell Theory

Theodor Schwann and Matthias Schleiden are usually credited with the formulation of key tenets of the cell theory. That the cell is the fundamental unit of life, and that all plants and animals are made of cells, each cell having a nucleus and a nucleolus (2). But it was not clear to them how new cells were made. They favored the idea that intracellular or extracellular matter crystallizes somehow into new cells (1, 2). Schleiden thought that new nuclei form without any relationship to preexisting ones. In essence, the views of Schwann and Schleiden on how new cells are made fell squarely into the realm of the spontaneous generation of life from nonliving matter, which was a popular theory at the time. Until Louis Pasteur put an end to spontaneous generation with his unambiguous *col de cygnet* (swan neck flask) experiment in 1859.

Explicit descriptions of cell division and the concept of new cells arising from preexisting ones came from Robert Remak (1, 2). Remak reached these conclusions from observations in multiple contexts, from red blood cells of

chicken embryos to frog eggs immediately after fertilization (1). The splitting of eggs after fertilization had been
observed by others before. But Remak's histological manip-
ulations allowed him to visualize the membrane of the
egg, and follow the origin of the embryonic cells from the
fertilized egg. The continuity of all the cells in the embryo
from one fertilized ancestor was now a settled issue. This
placed cell division and the cell cycle as the basis of how
multicellular organisms develop. Remak's data fit nicely
with Pasteur's that there is no spontaneous generation of
life. The notion that new cells arise only from cell division
was opposed by many, including Rudolf Virchow. But Virchow changed his tune later. Virchow's famous aphorism
omnis cellula e cellula (all cells arise only from preexisting
cells) encapsulated succinctly and popularized a key part
of the cell theory, a pillar of modern biology. Together with
the other pillar of modern biology, Darwin's theory of evolution, we arrive at a stunning and profound conclusion:
Starting with a single cell, all life that ever was, is, and
will be on this planet results from cell division. Spend a
moment to reflect on this. If you had any doubt that the
topics we will discuss in this book are important, now is
the time to put those doubts to rest.

1.1.2 Mitosis

By the mid-nineteenth century, the cell theory was
established, and microscopy was getting better. Several
individuals (including Remak) had documented distinct
stages of nuclear division, including nuclear elongation
in some cases, and nuclear dissolution in others. Wilhelm
Hofmeister noticed that the nuclei of plant cells dissolved,
but some nuclear material remained, in what he thought
were coagulates, which then segregated into the nuclei
of daughter cells. Although Hofmeister could not realize
the biological role of what he was seeing, his descriptions

were remarkably close to the stages of mitosis we recognize today (1). Scientists kept looking and looking under the microscope. They also processed and stained the cells with various techniques and dyes -all searching for crisp landmarks. Walther Flemming experimented with basic dyes and fixatives. To this day, pathologists use similar methods to look at cells in tissues. Flemming found that a nuclear substance, which was presumably acidic and negatively charged, stained very strongly with basic, positively charged dyes. He used the term chromatin (from the Greek "colored") to describe that substance. Flemming used salamander cells, because they were big and had big nuclei. As always, choosing the right experimental system for a research objective can reap enormous rewards.

Imagine you have a simple microscope, and you have figured out how to stain cells, with some nuclear substance being intensely stained. Assuming that it takes several hours to days for typical proliferating animal cells to divide, what would you expect to see? For the most part, not much. Each cell would be very similar to others and to itself, from its birth until it divides. Shortly before division, Flemming noticed that the colored nuclear substance was organized into threadlike structures (the Greek word for a thread is μίτος/mitos), which were then distributed into the daughter cells, in a process he called mitosis (Figure 1.1).

The nuclear threads/filaments were named chromosomes a few years later by Waldeyer. Unlike others that had also seen thread-like structures forming and segregating, Flemming was the first to discover their splitting during mitosis lengthways (1). Flemming imaged and documented the complete series of events during mitosis, a fundamental cellular process, in all its glory: first, chromosomes appear, becoming denser and more compact over time, while at the same time the nuclear membrane disappears in most animal and plant cells (prophase); second, the compact chromosomes reach an equilibrium position at an

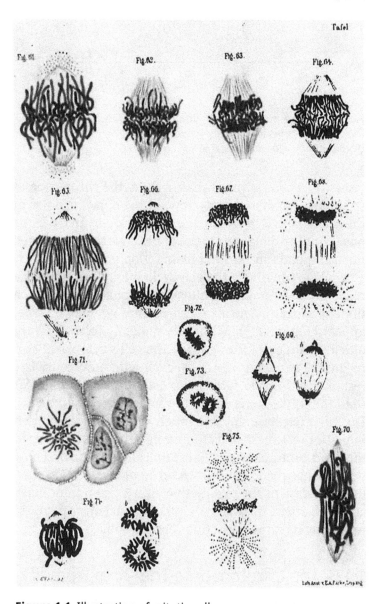

Figure 1.1 Illustration of mitotic cells.

Source: (3) Walther Flemming (1882), F.C.W. Vogel.

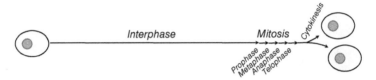

Figure 1.2 The duration of the interphase is much longer than the mitotic stages in the cell cycle. In most cells, interphase is devoid of any morphological landmarks.

"equator" position (metaphase); third, the chromosomes split lengthways and move toward the opposite poles of the cell (anaphase); fourth, the chromosomes arrive to the poles and the daughter nuclei appear, with reconstituted nuclear membrane around them (telophase).

Now we can divide the cell cycle into two phases. The relatively short phase of mitosis ("M" phase) starts when threads appear in the nucleus and ends when two nuclei appear (Figure 1.2). Mitosis is typically followed very quickly with cytokinesis when the cell's cytoplasm segregates to two daughter cells, each with a nucleus. The rest of the cell cycle, which in most cells lasts a lot longer than the M phase, is called the interphase (Figure 1.2). During interphase nothing much appears to be happening. The dramatic visual landmarks of mitosis are hard to miss and continue to guide research about when and how the nucleus divides. But it would be many decades later, in the middle of the twentieth century, when interphase landmarks were discovered. Until then, scientists followed what they had, the mysterious chromosomes.

1.1.3 The Chromosome Theory of Heredity

When the chromosomes were first seen, their role was unknown. That the chromosomes carry the genetic information would not be clearly formulated until 1903 from Walter Sutton, and at about the same time from

Theodor Boveri. The chromosome theory of heredity would be proven beyond doubt in the first decades of the twentieth century. Mendel's laws, which by 1900 were rediscovered, were abstract and could apply to any substance that carried the genetic information. As we will discuss later, the chromosomes do behave according to Mendel's laws. But what was the evidence that led to the idea that chromosomes carry the genetic information?

It was becoming more apparent that the nucleus, and not the cytoplasm, carried the genetic information. The unambiguous pieces of evidence for the role of the nucleus in carrying the genetic information came from fertilization experiments. Several individuals had already shown that the male and female parents make equal genetic contributions to the zygote, even though the egg is usually much bigger than the sperm. But these observations do not necessarily exclude cytoplasmic contributions to heredity. Until Eduard Strasburger showed in 1884 that during orchid fertilization, it is only the nucleus and not the cytoplasm (he coined the term) that is forced out at the end of the pollen tube into the embryo sac (1). Suppose both parents make equal genetic contributions to the zygote, but only the male parent's nucleus makes it into the embryo. In that case, the nucleus must carry the genetic information.

What nuclear component could be the carrier of genetic information? There was nothing better to look at than the chromosomes. If chromosomes carry the genetic information, then their continuity from generation to generation must be preserved. Again, the choice of the experimental system was critical. If you want to follow chromosomes, pick a system where you can track them easily. Edouard van Beneden chose *Ascaris* (a genus of roundworms) with only four (and sometimes just two), large, and easily distinguishable chromosomes, where he could follow the male and female-derived chromosomes in the zygote after fertilization. In 1883 he stated that each parent's chromosomes never mix and maintain their identity in the zygote and possibly in all the nuclei derived from it as the embryo develops (1). Boveri used the same system, *Ascaris univalens*, and extended van Beneden's observations to show that the number, morphology, and identity of each chromosome is maintained, division after division. Although in interphase the chromosomes are not visible, they are not destroyed. William Sutton made similar observations by looking at grasshopper cells, noting that each of their 11 pairs of chromosomes was morphologically distinct (4). In diploid organisms, two sets of chromosomes are brought together at fertilization, and they are inherited by all the cells of the new individual. It is only in the germinal cells through meiosis that the haploid state is generated again from the diploid state (1). Both Sutton and Boveri noted that their data about the constancy of chromosomes fit very nicely with Mendel's rules of inheritance: Genetic traits are stable from generation to generation, come in pairs (in diploid organisms), with each pair separating in gametes (during meiosis), but each pair segregating in gametes independently of how the other pairs segregate (because different chromosome pairs orient at random during division). All the evidence pointed to the direction that chromosomes and their behavior constitute the physical basis of Mendel's laws. Still, however, the evidence was mostly indirect. What was needed to unambiguously test the Sutton–Boveri chromosome theory was evidence that specific genetic traits localize to specific chromosomes.

Looking at a random gathering of people, say in a classroom, there are many traits one could identify. However, most show a broad distribution among individuals (e.g., height, weight, coloration of hair or skin, etc.), which could complicate their study. But one trait is unmistakably distributed in a precise pattern: a one-to-one male-to-female ratio. Could gender be linked to a particular chromosome? Nettie Stevens answered precisely that, looking at cells from *Tenebrio* beetles, which have a small Y chromosome that is easily distinguished from the larger X. Stevens saw that in *Tenebrio* (and as we know now in people too) the female was XX and the male XY (5). Somehow, Stevens' observation did not receive the attention it deserved. Thomas Hunt Morgan and his students developed *Drosophila* as a genetic system. They showed that specific mutant phenotypes are always associated with the transmission of particular chromosomes (6). Those triumphant and elegant *Drosophila* experiments have a special place in the history of genetics. They "sealed the deal" on the chromosome theory of heredity, and some are still standard laboratory exercises in undergraduate genetics courses.

Our brief overview of the cell and chromosome theories dealt with normal cells and organisms. The role of cell, nuclear, and chromosome division in disease was proposed early on, most notably from Theodor Boveri. In an astonishingly prophetic monograph published in 1914, accessible again in English (7), he described aberrant cell divisions associated with cancer, often leading to abnormal chromosomes in cancer cells. Many of Boveri's ideas were validated decades later, and they are now standard in cancer cell biology. Based on the evidence he had, Boveri proposed that *"The cells of even the most malignant tumors can be formed from normal tissue cells. The determinants of this abnormal behavior are to be found in the tumor cells themselves, not in their surroundings."* Indeed, every cell from a preexisting cell.

1.1.4 Deoxyribonucleic Acid (DNA)

With the chromosome theory in place, the frontier now shifted. To what chromosomes are made of and what exactly is the genetic material? To answer, it was biochemistry, not genetics, that was needed. In 1871, Friedrich Miescher was able to fractionate and purify a viscous, phosphorus-rich nuclear extract he called "nuclein" (now known as a nucleic acid). As early as 1884, Oscar Hertwig proposed that nuclein carries the genetic information (1), but this was speculation, not proof. Once chromosomes were established to have genetic information, scientists purified and studied their chemistry. It turns out that in addition to nucleic acid (mostly DNA and some RNA), chromosomes have a lot of protein. The precise ratios vary among different cell types, but DNA is rarely even the majority in chromosomes. Generally, eukaryotic chromosomes are 30–50% DNA, 1–10% RNA, and 50–65% proteins (histones and nonhistones). Furthermore, looking at each of these fractions' composition, nucleic acids were found to be the least chemically diverse group compared to proteins. The genetic material must account for the incredible diversity of life. It is only natural that nearly all scientists until the middle of the twentieth century thought that biological diversity should be matched by a chemical one. Hence, they placed their bets on chromosomal proteins being the physical basis of heredity, not nucleic acids.

To do biochemistry, you have to work with nonliving matter, to fractionate and purify it. But cells are the unit of life and they are only generated through cell division. One cannot make a cell in a test tube from inanimate matter. So how can you do biochemistry to show that something is the genetic material? The door opened with Griffith in 1928, and his "transforming principle" experiment (8). He found that a substance from nonliving cells could be transferred to and change the traits of living cells. It is the genetic material (the genotype) that specifies a range of possible traits (the phenotype). Griffith's experiment showed that the genetic material need not be living itself. Oswald Avery and colleagues realized that they could separate the transforming but inanimate matter in different fractions, and repeat Griffith's experiment using the individual, purified fractions. The DNA fraction was necessary and sufficient to change the phenotype (9). The experiment was simple, lucid, and of the utmost significance. It defined the physical basis of heredity. Yet, it did not receive the universal attention it deserved (including by the Nobel Prize committee). Some scientists did notice, including Maurice Wilkins, who was convinced to continue his work on DNA using X-ray crystallography. Wilkins would eventually share the Nobel Prize with Crick and Watson for determining the structure of DNA. Probably in one of the most famous sentences in the scientific literature, Watson and Crick noted at the end of their paper (10): *"it has not escaped our notice that the specific pairing we have postulated immediately suggests a possible copying mechanism for the genetic material."* What followed is history and well-covered elsewhere. What concerns us at this point is the pattern of DNA synthesis in the cell cycle. As it turns out, DNA synthesis is a key landmark in the cycle.

Various methods can monitor the synthesis of macromolecules in the cell. For example, feeding the cells with a labeled precursor and then following its incorporation in a macromolecule. The task becomes easy when the labeled precursor is uniquely incorporated into the macromolecule of interest. This is the case for tritiated thymidine in DNA. Around the same time as the Watson and Crick paper, these kinds of experiments showed that in animal and plant cells, DNA was synthesized during a fraction of the interphase and that it led to a precise doubling of DNA. The period of DNA synthesis is called the S phase. The interphase portions ("gaps") before and after S phase are termed G1, and G2, respectively (11). G1, S, and G2 are followed by mitosis, M. So now we have another landmark in the cell cycle. Unlike the morphological landmark of mitosis, however, the S phase is a molecular one. In most cases, one cannot tell if a cell is in the S phase by just looking at it. Nonetheless, with the proper molecular techniques, the start and finish of the S phase are easily defined. Defining where G2 ends and M starts is not as straightforward. The appearance of compacted chromosomes defining the start of the M phase is a morphological landmark, open to different standards and techniques to visualize the chromosomes.

1.1.5 Cell Cycles Come in Many Flavors

The G1 → S → G2 → M description of the cell cycle phases is universally accepted and used. As we will discuss in detail later, each phase's duration in absolute terms and relative to each other can vary over a broad range across different cell types and for the same cells proliferating under different conditions. The canonical G1 → S → G2 → M scheme applies to most human somatic cells, with the cell cycle ending with cytokinesis (Figure 1.3). However, the canonical scheme is far from universal.

In the very rapid cell cycles of the embryo after fertilization, G1 and G2 may be absent. These cycles start from a very large fertilized egg, and the only tasks are to

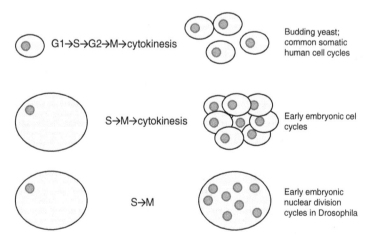

Figure 1.3 Embryonic cell cycles are usually devoid of gap phases.

duplicate and segregate the genome rapidly. There is no need to duplicate the cytoplasm because the egg is large enough to provide the necessary cytoplasm to as many as thousands of daughter cells. For example, embryonic *Xenopus* cell cycles starting from the huge fertilized egg are successive S → M → cytokinesis cycles (each lasting about 30 minutes), with a proportional reduction in cell size with every division. In *Drosophila* embryos, the scheme is pushed even further to a simple S → M (each lasting only eight minutes), with no time for cytokinesis (Figure 1.3). After about 13 such embryonic cycles, the result is a large cell with the same size as the fertilized egg, but with thousands of nuclei. At that point, the rapid divisions stop, and a cell membrane forms around each nucleus. Until then, the *Drosophila* embryonic cycles are technically *nuclear* division cycles, not *cellular* ones. After these embryonic divisions, subsequent cell cycles are slower and follow the canonical (i.e., G1 → S → G2 → M) scheme.

In most cases, S and M phases are temporally separated, with genome duplication (S) completed before its segregation (M). However, this temporal spacing may not hold in some rapid embryonic cycles, with a new S phase starting before the previous M phase is over. The timing of cytokinesis also varies among different cell types and organisms. If cytokinesis is fast, as it is in human cells, it usually happens quickly after telophase in the M phase. In fission yeast, cytokinesis is slow. It starts after M, but in rich nutrients, fission yeast enters the next S phase soon after M is completed (therefore lacking a G1 phase). As a result, cytokinesis ends even after S is completed, in the G2 before the next M. Why is this happening? It seems strange until one realizes that fission yeast spends most of its life cycle in the haploid state, with only one copy of each of its chromosomes. In rich nutrients, fission yeast has minimized or eliminated G1. Still, it spends a considerable amount of time in G2 with a duplicated genome, which affords more DNA repair possibilities in case of damage.

On the other hand, budding yeast prefers to proliferate in the diploid state, and its cell cycle shows a typical G1 phase. However, budding yeast has tiny chromosomes, challenging to see, and it does not have a clearly defined G2 phase. Despite these eccentricities, both yeasts have been and continue to be prime cell cycle model systems because of their unique experimental properties, which will become apparent in the next chapter when we discuss their cell growth pattern.

Regardless of the different ways the cell cycle phases are arranged in the above examples, the genome complement is maintained in all cases. It is accurately duplicated and segregated in subsequent generations. What happens if there is no M or S phase, or when they exist with a twist? We will try to answer these questions briefly before ending this chapter.

Several cell types go through successive S phases with no M phase, from insect larvae to our liver cells. These endoreduplication cycles increase the copy number of the genome, which likely enables overproduction of gene products. In the extreme case of salivary gland cells in *Drosophila* and other insects, ten rounds of endoreduplication yield cells with 1024 copies of each chromosome! Furthermore, these copies align with each other, in perfect register, making them easily visible. Such polytene chromosomes are an excellent cytogenetic resource, with clear, visible markers along each chromosome (Figure 1.4). In many cases, they have enabled fly researchers to localize mutations physically.

Sexually reproducing diploid organisms start from the zygote with two nearly identical copies of each chromosome, one from each parent. The vast majority of the progeny, in somatic cells, is generated by the canonical mitotic cell cycles we described. Each chromosome from

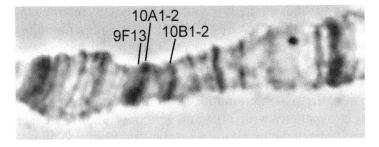

Figure 1.4 Polytene chromosomes from the salivary gland of a fruit fly larva. These chromosomes show many distinct, darkly staining bands (indicated by the labels in this image), providing clear physical chromosomal landmarks.

Source: (12) Vatolina et al. (2011), PUBLIC LIBRARY OF SCIENCE (PLOS), CC BY.

both parents is duplicated in the S phase and represented by two sister double-stranded DNA molecules, or chromatids, in the G2 phase. The sister chromatids of each of the parental chromosomes then segregate lengthways in the M phase. Hence, at the end of a mitotic cell cycle, each daughter cell receives an exact diploid set of chromosomes. But in the germline, how do these organisms generate their haploid gametes? The key difference is that in the first meiotic division, it is not the sister chromatids that segregate lengthways after DNA replication. Instead, the homologs from each parent align together in metaphase. They rarely do so in the mitotic cell cycle. In anaphase, the two identical sister chromatids of each chromosome do not separate, but the homologs from each parent do. Hence, although the products of the first meiotic division (a.k.a. reduction division) have the same genome content as the original cell, they do not carry two homologs from each parent anymore. Just a duplicated homolog from one parent only. Which parent contributes each homolog is random, depending on how the duplicated homologs oriented themselves in metaphase, accounting for Mendel's law of independent assortment of different traits. This first meiotic division is beautiful and extremely important for sexually reproducing organisms like ourselves. The forces that align homologs but keep sister chromatids together are special and not seen in the typical mitotic cell cycles. The second meiotic division is just like a mitotic cell cycle, but without an S phase. Then, the sister chromatids separate in anaphase, giving rise to gametes with a haploid genome content and each unduplicated chromosome originating from one parent, but not both.

As spectacular and important meiosis is, we will not cover it in more detail. Likewise, we will not discuss some unusual and exotic eukaryotic cell cycles, for example, among protists. Lastly, there are several key differences in prokaryotic vs. eukaryotic cell cycles, perhaps most notably the initiation of multiple DNA replication rounds in one cell cycle. In rich nutrients, bacteria take advantage of the fact that they can double their cytoplasm faster than they can duplicate their genome. This is a short book, and we will not discuss much about bacterial cell cycles. Nonetheless, we will close with the observation that as more studies are reported, the differences in the mechanisms of cell division between prokaryotes and eukaryotes are getting less pronounced, not more.

2 Cell Growth and Division

- How do we measure cell growth?
- What is the relationship between cell growth and division?
- When and how do cells grow in the cell cycle?

2.1 Balanced Growth and Cell Proliferation

As the unit of life, the cell must carry out all the processes associated with life. Cells are open thermodynamic systems, exchanging both matter and energy with their environment. The entire enterprise of synthesizing a cell encompasses many more processes than those needed to duplicate and segregate the genome. Furthermore, the rate at which cells can divide usually depends not on how fast they can replicate their genome but on how fast they can synthesize everything else that goes into a newborn cell. If *"The dream of every cell is to become two cells* (Francois Jacob)," it is cell growth that makes those dreams come true.

Imagine an environment where cells have constant nutrients and growth factors, and toxic products do not accumulate. When the number of cells is measured over

Two from One: A Short Introduction to Cell Division Mechanisms,
First Edition. Michael Polymenis.
© 2023 John Wiley & Sons Ltd. Published 2023 by John Wiley & Sons Ltd.

successive cell cycles in such an environment, cell growth and division appear balanced. General cellular properties (e.g., cell size and composition) and cell proliferation rate remain constant. In these situations, cell proliferation behaves like a first-order autocatalytic reaction. The change in the number of cells (N) over time (t) can be described with a simple equation: $dN/dt = kN$. The proportionality constant, k, is the specific proliferation rate constant. It solely reflects cells' intrinsic properties and their ability to grow and divide in that particular environment. The apparent balance of growth and division means that the time it takes to double the number of cells is the same as the amount of time it takes to duplicate every cell component, from chromosomes to individual protein and RNA molecules. A plot of the logarithm of the number of cells over time will be a straight line, with a slope equal to k (Figure 2.1).

This is what is meant with experiments in "log-phase" cultures. It is essential to recognize that these relationships

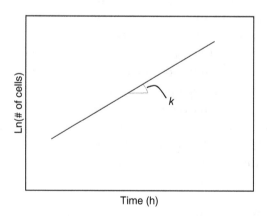

Time (h)

Figure 2.1 Measuring the rate of cell proliferation. In exponentially proliferating cell cultures, plotting the natural logarithm (Ln) of cell numbers (y-axis), over time (in hours, h; x-axis) should yield a straight line, with a slope (k) equal to the specific proliferation rate constant.

represent phenomenological population averages. For example, the age structure of the population is not taken into account. The cell cycles of newborn cells may be different from that of their mothers. Furthermore, most microbes in nature are unlikely to encounter constant conditions of nutrients and growth factors. Nonetheless, these simple observations led to profound insights into the properties of proliferating cell populations and underscore the balanced coupling of cell growth and division (13). It is also only in conditions of balanced growth and division that all cellular parameters at the population level remain constant, enabling reproducibility of various studies from experiment to experiment, done by different investigators at different times and places.

Even nondividing, quiescent cells balance their low level of cellular biosynthesis, necessary to sustain life and cellular function, with a low level of or no cell division (Figure 2.2). Unless they are terminally differentiated, quiescent cells

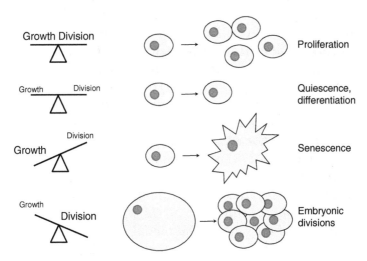

Figure 2.2 Possible outcomes of the relation between cell growth and division.

Source: Figure adapted from Polymenis and Kennedy (14).

maintain the potential to divide. In contrast, senescent cells keep growing without ever dividing, growth and division are not balanced, and they will get progressively bigger and die. In embryonic cell cycles, we have already encountered the opposite scenario, where successive divisions of the large zygote without cell growth eventually lead to thousands of much smaller cells. But embryonic cell cycles are unique. They rely on preexisting cell growth, which took place in the oocyte before fertilization. In all other continuously proliferating cells, cell growth and division are balanced with each other (Figure 2.2, top). How that happens remains mostly mysterious. Even general processes that could mediate this coordination between cell growth and division are not adequately defined, or they are not considered part of the cell cycle.

Suppose you want to find the current definition of any cellular process and what other processes are part of or interact with it. In that case, arguably, the best place to look for that kind of information is gene ontology (http://geneontology.org/). The stated goal of this great collaborative effort is to offer "*The network of biological classes describing the current best representation of the 'universe' of biology.*" Indeed, the term "cell cycle" (GO:0007049) describes "*The progression of biochemical and morphological phases and events that occur in a cell during successive cell replication or nuclear replication events.*" The related term "mitotic cell cycle" (GO:0000278) is defined as "*Progression through the phases of the mitotic cell cycle, the most common eukaryotic cell cycle, which canonically comprises four successive phases called G1, S, G2, and M and includes replication of the genome and the subsequent segregation of chromosomes into daughter cells.*" Each process has "child" terms to indicate related processes that are part of the "parent" term. For example, in the budding yeast ontologies, there are more than 60 child ontology terms of the "cell cycle" and "mitotic cell cycle" terms, many dealing with chromosome duplication and segregation or the cytoskeletal rearrangements necessary for cell division (15). But no ontology describes a process of how cell growth is linked to cell division. How could that be? To appreciate the problem, we will deal with a series of questions. They have to do with the different meanings of cell growth, ways to measure growth, the settings in which it is measured, and the models that attempt to account for the measurements during the cell cycle.

2.2 Measures of Cell Growth

By cell growth, we do not mean the increase in cell numbers over time. We mean some parameter that reflects the making of new cellular matter in individual cells. There are several such parameters.

The volume of each cell is commonly used to gauge cell growth. It encompasses everything in a cell, large molecules and small ones, including water. Volume is measured easily from live cells in solution, using instruments that rely on the Coulter principle. As cells are pulled through an aperture in an electrolyte solution concurrently with an electric current, they impede the electrical flow. The magnitude of the impedance is proportional to the size of each cell (Figure 2.3). The output of these measurements

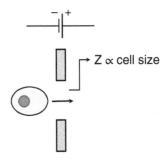

Figure 2.3 The Coulter principle measures changes in impedance (Z), which are proportional to the volume of a cell that passes through the aperture.

is a cell size distribution of the sample. The accuracy and wide availability of the instruments (they are used extensively in clinical labs to count cell numbers in the blood) explain the approach's popularity.

Measuring mass instead of volume is not trivial. It requires putting a cell on some scale with precision and accuracy to a few hundredths of a picogram. Recently, the development of microfluidic resonators allowed measurements of the resonant frequency of live cells, which is a function of the cell's mass (16, 17). This methodology measures the buoyant mass of cells and surpasses all other approaches in precision and accuracy. Getting precise and accurate measurements becomes critical when we will examine patterns of cell growth in the cell cycle.

Cell size may be a holistic reflection of cell growth, but it includes compounds that are not necessarily reflections of biosynthesis. "Dry" mass measurements exclude water but include all other compounds, large and small. Note that buoyant mass is dependent on the amount of biomass in the cell, most of which is denser than water, and so is analogous to the dry mass of the cell. Various biophysical methods that rely on optical properties have also been used to measure the dry mass of live cells (17). Starting with interference microscopy in the 1950s, the relationship between mass and refractive index has been used to convert cell refractive index distributions to dry mass measurements. When applied to nonliving fixed cells, interferometry measurements will mostly reflect the macromolecular mass. However, the outcome will depend on how the cells were fixed because some low molecular weight compounds may not be lost during fixation.

It is already becoming apparent that, depending on how one measures growth, the answers can vary. Indeed, volume and dry mass measurements in the cell cycle may not match. Cell size is all-encompassing, but it may change for reasons unrelated to making new cellular components.

But one type of measurement is not necessarily better than the other. Mass changes may better reflect biosynthesis. Volume determines the concentration of various compounds within the cell, although the numerous organelle compartments in eukaryotic cells make estimates of effective concentrations far from straightforward. On the other hand, dry mass includes the low molecular weight metabolite fraction, a quarter of the total dry mass in some cells. If we measure biosynthesis, should the low molecular weight pools be counted? Such difficult-to-answer questions illustrate why it is a combination of experimental approaches needed to get an accurate idea of cell growth (Figure 2.4). This point will become more evident when one looks at macromolecular synthesis.

Before examining a macromolecule's synthesis to gauge cell growth, we must consider what a cell is made of. Naturally, numbers vary for different kinds of cells. For example, the carbohydrate fraction of plant cells is very high, needed to construct their cell walls. For most dividing cells, proteins are the most abundant class. For example, in fast-dividing budding yeast, proteins are 46% of dry mass, carbohydrates 32%, lipids and RNA are at 8% each, and

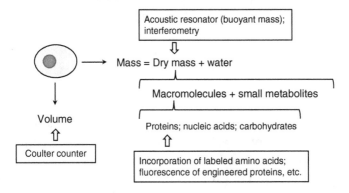

Figure 2.4 Different aspects of cell growth are captured by different techniques.

DNA only 0.5% (18). The abundance of proteins explains why protein synthesis is considered a prime indicator of cell growth. Total protein synthesis is also easy to quantify by measuring the incorporation of labeled amino acids into newly synthesized proteins (Figure 2.4). But such analytical approaches demand material from populations of cells, and they cannot be used on the same living cell continuously.

An alternative is to engineer the expression of one or more fluorescent proteins to monitor protein synthesis in single living cells. Fluorescence is then used as a readout of protein synthesis. The approach can be automated, tracking under the microscope many cells simultaneously. The methodology has the advantage that single cells can be monitored continuously, as they grow and divide. Nonetheless, it is an indirect method, relying on several steps, each of which has to be controlled, from the segmentation algorithms used to identify each cell's boundaries to how the true fluorescence signal is separated from imaging artifacts. It is also vital to ensure that cellular physiology has not changed and, if a protein has been engineered to fluoresce, it has the same properties as the native protein. Lastly, extrapolating from the pattern of synthesis of one protein to bulk protein synthesis or even to overall cell growth (cell size, dry mass, etc.) can be complicated and not without caveats.

2.3 The Relationship Between Cell Growth and Division

The fact that dividing cells balance their growth with their division does not reveal which of the two processes controls the other. The genetic information in the chromosomes dictates the kind of proteins the cell will have. It was also known since the time of Oscar Hertwig and Theodor Boveri that any given type of cell somehow maintains

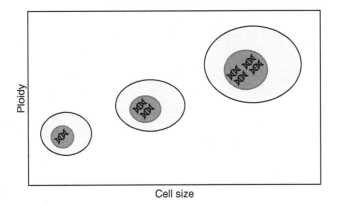

Cell size

Figure 2.5 Increasing ploidy also proportionately increases cell size.

a characteristic constant proportionality between the volume of its nucleus and cytoplasm, the N:C ratio. When the nucleus gets bigger (e.g., by increasing the ploidy) so does the cytoplasm (Figure 2.5).

Even in the face of a severe reduction in their cytoplasmic components, as long as cells have a normal nucleus and chromosomes, they will eventually generate the rest of their cytoplasm and segregate it into daughter cells. This is why the duplication and segregation of cytoplasmic components have traditionally received much less attention in cell cycle studies than the genome's replication and segregation. But, as it turns out, having the right amount of cytoplasm determines if and when cells will divide.

Early hints to that effect came from cellular amputation experiments. The rationale is straightforward: To test if the cytoplasm exerts some control over nuclear division, remove some cytoplasm and see what happens. The "patients" of these cellular surgeries were the large *Amoeba proteus* cells. Usually, they divide every 24 hours under standard laboratory conditions. Unless a portion of their cytoplasm is amputated. In that case, the amoeba will not divide until it produces more cytoplasm. Hartmann, in the 1920s, repeated these cytoplasmic amputations on the same cell for months! That cell did not divide, while its sister cell (which arose from the same original division that gave rise to the amputee cell) divided 65 times. Was the inhibition of division down to a fundamental relationship between cell size or cytoplasmic mass and control of cell division? Prescott reproduced these experiments in the 1950s and noted that the interpretation is not straightforward. Cytoplasmic amputation in a growing amoeba delays cell division and reduces the rate at which the cell grows in size and the volume of its nucleus (19). As dramatic as the amputation experiments were, they could not be extended to conclusive ends. More penetrating experiments would be needed. Enter a genetic dissection of the relationship between cell growth and division.

To do genetics, you need mutants, which you can then study and observe their phenotypes. But how do you get mutants of an essential cellular process that, by definition, would be dead and not very amenable to follow-up experiments? Borrowing a strategy used to study bacterial viruses, Lee Hartwell attacked this problem by generating conditional, temperature-sensitive mutants in budding yeast (20). These landmark studies propelled our understanding of the cell cycle, and they will be described in more detail later. With those mutants at hand, Hartwell also probed the relationship between growth and division. He asked a simple question: If cell division is blocked, do the cells continue to grow? The answer was yes. Cells arrested at different points in the cell cycle continued increasing and ballooning in size before eventually dying. On the other hand, stopping cell growth also blocked successive cell divisions. Years later, with the key discovery by Mike Hall and others of the Tor kinase, the "master regulator" of multiple anabolic pathways, including protein synthesis, it was clearly shown that upon loss of Tor activity, cells arrest in early G1 with a small size (21). In some instances, the cell cycle machinery does trigger specific "growth" pathways, especially in lipid and carbohydrate mobilization, presumably to provide resources necessary for cell division (22, 23). Nonetheless, by and large, growth controls division and not the other way around.

2.4 Patterns of Growth in the Cell Cycle

Since cell growth controls cell division, how do cells know when they have grown enough to divide? Before we explore this question, we must look at the pattern of growth in the cell cycle. Suppose you observe newborn cells and measure their growth somehow as they progress synchronously in the cell cycle. The first question you could answer is if the cells grow all the time in the cell cycle, or perhaps in a stepwise fashion, with little or no growth at some phases. Second, when cells grow, and some aspect of growth is plotted over time, one could estimate how they grow. Answering these questions is not trivial, and the answers also depend on the experimental system in question. The answers have implications about how cells couple their growth with their division and what kind of intrinsic mechanisms they use, if any, to monitor their growth.

We will mostly focus on single cells as they progress in the cell cycle. We have already seen that when cell growth and division are balanced, cell numbers increase in an exponential pattern. That pattern may be an emergent population average, not necessarily reflecting how individual cells grow and divide under these conditions. If cells grow during their entire cell cycle, assuming that the "new" cell mass can immediately contribute to new biosynthesis in the same way as the "old" cell mass, the rate of adding new mass would continuously increase. Hence, the expected pattern of cell growth at the single-cell level

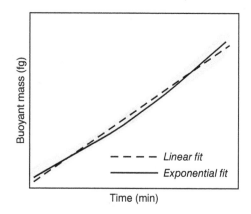

Figure 2.6 Plots of exponential and linear growth over one doubling are very similar. Experimental data, dispersed within the gray space, for *B. subtilis* over roughly one cell division.

Source: Figure adapted with data from Godin et al. (16).

will also be exponential. Put another way, in exponential growth, bigger cells grow in mass faster than smaller ones do. Another scenario is linear growth, where the amount of mass added to the initial cell is constant. Linear growth implies that as cells get bigger, they become less efficient in producing more mass. A significant problem in distinguishing exponential vs. linear patterns of growth is that the plots look very similar (see Figure 2.6). Hence, accurate and precise measurements become essential.

2.4.1 Amoeba Cell Growth

Prescott measured the weight of individual *Amoeba* cells in water (corresponding to dry mass), cell volume, and protein content from birth to division (24). In a roughly 24-hour cell cycle that ends with mitosis (which takes up the last 30 minutes or so), all parameters had the same pattern, with no increase in the last four hours (24). As *Amoeba* cells increase in size, their growth rate drops (25), arguing for nonexponential growth.

2.4.2 Fission Yeast Growth

The growth pattern of individual fission yeast cells in the cell cycle has been monitored since the 1940s. Fission yeast cells grow lengthwise (Figure 2.7a). This is a valuable property because their cell cycle position can be inferred by simply measuring their length. Fission yeast measurements of growth in the cell cycle offer an excellent example of the different answers one gets, depending on what aspect of "growth" is measured. Dry mass appears to increase in a linear pattern. In contrast, volume increases exponentially until three quarters into the cell cycle, when the septum begins to form, and then it remains constant (26). Remember that dry mass includes both macromolecules and low molecular weight compounds. It turns out that protein, RNA, and carbohydrate content all increase exponentially in the fission yeast cell cycle. How can it all make sense? The discrepancy between mass and volume argues that there are density changes in the cell cycle.

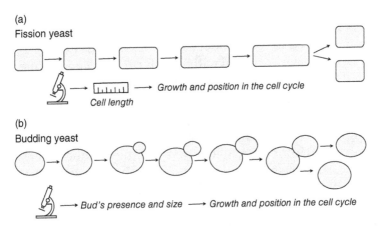

Figure 2.7 The ways (a) fission and (b) budding yeast cells grow in the cell cycle provide morphological "metrics" of cell growth and position in the cell cycle, accessible in live cells with simple and inexpensive microscopes.

Furthermore, since dry mass increases linearly, but the macromolecular fraction exponentially, the low molecular weight fraction must not do so, likely accounting for the different patterns.

2.4.3 Budding Yeast Growth

Budding yeast's growth in the cell cycle offers unique experimental advantages, which were instrumental in Lee Hartwell's landmark genetic dissection of the cell cycle. You cannot just look at the cells in most cell types and know if they have started replicating their DNA. From the outside, a mammalian cell in G1 appears the same as a cell in S or G2. However, budding yeast announces to the world when it starts replicating its DNA by making a tiny bud on its surface. Then, most of the new mass the cell generates is directed in the bud (Figure 2.7b). The bud continues to increase in size and becomes the new daughter cell at the end of the cell cycle. The bud (absent or present) and its size enabled Lee Hartwell to differentiate his conditional mutants' morphology. Mutants proliferating asynchronously at their permissive temperature (25 °C) were shifted to their nonpermissive temperature (37 °C). Cells that were arrested with a uniform morphology (e.g., with buds of similar size) must have reached a point in the cell cycle that they could not complete. Such mutants with a uniform arrest phenotype were defined as cell division cycle (cdc) mutants. In contrast, mutants arrested with many different morphologies, reflecting a random arrest not specific to any particular point in the cell cycle, likely affected some other general process, not related to the cell cycle itself.

After this short detour (more on the cdc screen later), let's look at measurements of growth in the budding yeast cell cycle. There are now large data sets measuring several cell growth parameters, in synchronous populations and single cells. The growth pattern of newborn daughter cells appears exponential. This conclusion is supported both by population-based and single-cell studies. The mean cell size from populations of daughter cells progressing synchronously in the G1 phase was measured continuously with a modified Coulter counter. Such continuous data sets afford high resolution, enough to distinguish between linear vs. exponential growth. The fit was better as an exponential function ($R^2 = 0.972$) than a linear one ($R^2 = 0.932$) (27). Single-cell studies used the constitutive, strong expression of a fluorescent protein as a proxy for total protein content (28). When the total fluorescence per cell was quantified, the data fit better to an exponential pattern of growth (28). Combining measurements from a microfluidic resonator and a Coulter counter suggested that the cell density of daughter cells is not constant in the cell cycle, increasing noticeably before bud formation, arguing for discrepancies between volume and mass measurements (29). Lastly, although newborn daughter cells probably grow exponentially, mother cells do not. Budding yeast mother cells will divide about 25–30 times before they die. They get bigger with every successive cell division, but their increase in size from one division to the next is linear (30). Growth and division are not well coordinated in old mother cells. As they divide more times, protein synthesis becomes dysregulated and inefficient (31). It becomes apparent that in a population proliferating asynchronously, in which at any given time about half of the cells are newborn daughters, there are subpopulations with different growth patterns in the cell cycle.

2.4.4 Mammalian Cell Growth

Early studies found that in cultured mammalian cells, dry mass, protein synthesis, and volume all increased continuously in the cell cycle, with the possible exception during mitosis. The data lacked the precision needed to distinguish between exponential vs. linear growth (26). Volume measurements have shown linear growth for rat Schwann cells (32). On the other hand, mouse lymphoblast cells show exponential growth, but with a varying rate constant (33). Following mouse lymphoblasts with a microfluidic resonator showed an increase in growth rate during the G1 phase, but after the G1/S transition, the growth rate increases more slowly (34). Furthermore, newborn lymphoblasts entered G1 with different growth rates, but by the time these cells entered the S phase, they had reached similar growth rates even though their size was different. These data argue for some critical growth rate threshold that mammalian cells must reach before initiating DNA replication. It also appears that mammalian cells swell in mitosis, their density decreasing because their volume increases to a greater extent than their mass (35).

The above examples are not exhaustive, and for other cell types and organisms, different patterns of growth may be observed in the cell cycle. However, they are sufficient to set the stage for the next section, looking at general models of coupling growth with division.

2.5 Sizers vs. Adders

In proliferating cell populations, a manifestation of balanced growth and division is the constancy of all basic cellular parameters. An example of such homeostasis is cell size. For any given cell type and growth condition, the mean cell size in the population stays within some characteristic limits. Cells born abnormally large or small quickly return to the normal range within one or a few cell divisions. Especially for animal cells, there has been intense debate about whether size homeostasis is a passive or active process. In the passive scenario, cell size control emerges indirectly, without cells sensing how big they are. On the other hand, if the size is actively controlled, cells must have size-sensing mechanisms and then use that information to make the necessary adjustments in the cell cycle to maintain the correct size as they divide.

Size homeostasis is closely linked to the pattern of growth in the cell cycle. If growth in the cell cycle increases as a linear function, then homeostasis could be maintained without the need for an active, size-sensing mechanism (32). To illustrate, let us look at a hypothetical scenario where the average cell size at birth in a population is kept around five arbitrary mass units (Figure 2.8). Now imagine two cells, one abnormally large (at nine units) and one abnormally small (at one unit) produced from an unequal division. How would they return to the normal cell size? If growth is linear, then a constant amount of mass would be added per cell cycle (say five units per cell cycle), regardless of their starting birth size. At the end of the first division, the large cell would be $9 + 5 = 14$ units, and it would generate two cells of seven units each. At the end of the second division, each of these seven-unit daughters will add five mass units and generate two granddaughters

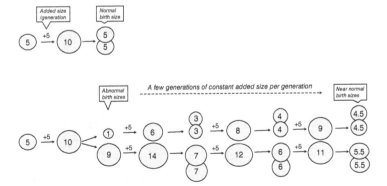

Figure 2.8 A constant added mass per generation could account for size homeostasis because within a few generations the population would return to its normal range of birth sizes.

of six units each. Consider one more generation, and the great-granddaughters would be 5.5 units each, pretty close to the target average of five units. Likewise, the abnormally small cell will generate two daughters of three units each at its first division, then four-unit granddaughters, and 4.5 unit-great granddaughters. If growth increases linearly in the cell cycle, abnormal sizes return to normal within a few generations (Figure 2.8). Size homeostasis is achieved without any need for an active mechanism that senses size and adjusts cell cycle progression accordingly. The above explains why so much effort has gone into defining whether cell growth increases linearly or exponentially in the cell cycle. Despite the technical difficulties, buoy-ant mass measurements in proliferating bacteria, budding yeast, and mouse lymphocytes are inconsistent with simple linear growth (16). A clear conclusion was that heavier cells grow faster than lighter ones, arguing for active mech-anisms that balance growth with division.

There are three general models that could account for the observed size homeostasis. First, "sizer" models, where cells can sense their size, and know how big they have to be before they can complete a cell cycle transition. Second, "adder" models, where cells know how to add a fixed amount of mass per cell division, independent of their size at birth. Third, "timer" models where cells divide after a certain amount of time has elapsed since their birth, following an internal clock. There has been little experimental support for timer mechanisms (a very weak timer may operate in *C. crescentus*), and we will not discuss the timer scenario further. Plotting how much cells increase in size in one cell cycle (the difference bet-ween size at birth and size at division; on the y-axis), as a function of size at birth (on the x-axis) can, in theory, distinguish these models. In ideal situations, the linear regression slope would be -1 for sizers, and 0 for adders (Figure 2.9a).

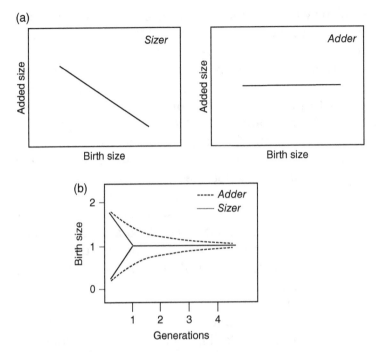

Figure 2.9 Sizers and adders. (a) Plotting added size (the difference between size at birth and size at division; *y*-axis), as a function of size at birth (on the *x*-axis) distinguishes a sizer (left; negative slope) from an adder (right; slope close to zero). (b) Outlier cells converge on the steady-state value (shown as 1) of relative birth size at different speeds depending on the model. The division control model is shown with solid (sizer) and dashed (adder) lines. Two outlier cells, one small (0.25 times the steady-state birth size of 1) and one large (1.75 times the steady-state birth size of 1), are shown in each case.

Source: Adapted from Vuaridel-Thurre et al. (36).

The models make different predictions about how many generations it will take for cells that are born too big or too small to return to the normal size range (see Figure 2.9b). An adder situation will be like the scenario described above, taking a few generations. On the other hand, if there is a sizer, an abnormally small cell will keep growing without completing a cell cycle transition until it has reached a certain size threshold. Within one generation, size homeostasis will be achieved (36).

The main point here is that a sizer, but not an adder, will actively adjust cell cycle progression. Sizers are thought to exist in both fission and budding yeast. Remember that in both organisms, cell size is used as a proxy for cell cycle position. This property has been of tremendous value in cell cycle studies. In rich nutrients, the major sizer in fission yeast operates at the G2/M transition. If a fission yeast cell is born with a shorter length, it elongates more before entering mitosis and vice versa. Plotting the increase in size in one cell cycle as a function of size at birth for fission yeast gives a strong negative slope of −0.76, close to the theoretical value of −1 expected for a perfect sizer (37).

In budding yeast, the situation is more complicated. Daughter cells are smaller than their mothers, and they have a longer cell cycle (Figure 2.10). They stay longer in the G1 phase and do not initiate DNA replication until they grow enough, reaching a "critical size" threshold. The critical size is not fixed and changes with nutrients (smaller in poorer ones) and ploidy (larger with higher ploidy). But for every given strain and growth condition, critical size is a characteristic and highly reproducible parameter.

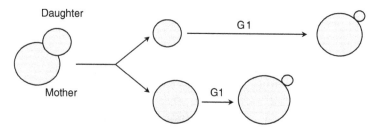

Figure 2.10 Daughter budding yeast cells are born smaller and stay longer in G1 until they reach a critical size threshold.

Overall, the inverse correlation between how big a cell is when it is born and how long it spends in the G1 phase of the cell cycle is indicative of a sizer. Indeed, plots of added size per G1 for daughter cells have a slope of −0.7 (28), consistent with a sizer, albeit an imperfect one. In contrast, for mother cells, the slope is close to 0, consistent with an adder mechanism. Furthermore, even in daughter cells, the sizer operates mostly in the G1 phase and not in subsequent phases (28). As a result, if one does not deconvolute all these different contributions in budding yeast, an aggregate adder behavior emerges (38). All this says is that growth patterns in the cell cycle need not be uniform and the same for all cells, especially in populations with complex age structures and cell cycles.

In other organisms, most of the data come from bacteria. Adder mechanisms predominate (37). In animal cells, ex vivo tissue culture studies also support adder phenomena. But in one recent in vivo study of mouse epidermis, a G1 sizer was reported, similar to the one in budding yeast daughter cells (39). As more and more data becomes available from different tissues in vivo, the patterns will become more apparent.

We will close this chapter with the realization that much about these patterns of growth in the cell cycle revolves around regression analyses and best fits (linear, exponential, and the magnitude and sign of the slopes on these graphs). The approaches are necessary, but they do not reveal the molecular underpinnings. In adder mechanisms, how do cells know they have added a fixed amount of mass? What properties could a sizer molecule have, and are there examples? If different mechanisms operate in different cell cycle phases, precise measurements of growth patterns need to be superimposed with exact measurements of the duration of specific cell cycle phases. How is this done? Some answers to those questions will be given in

3

Assaying Cell Cycle Progression

- How do we keep a score of where cells are in the cell cycle and how long each cell cycle phase lasts?
- How does the duration of cell cycle phases change in different conditions?
- How do we get populations of cells that progress synchronously in the cell cycle?

3.1 Measuring Cell Cycle Phases

Following cell cycle kinetics is a lot like being on a road trip. It is not enough to know how long the whole trip will take. One must also follow the mileposts, plan for stops, transitions in terrain, and know how long each segment of the journey will be. For the same reasons, we need to know when each phase of the cell cycle begins and ends and how long it lasts, both in absolute terms and relative to the whole cell cycle.

3.1.1 Single-Cell Imaging

There is a dearth of natural "mileposts" and landmarks along the cell cycle visible from the outside. In most animal and plant cells, mitosis is the only cell cycle phase that can be visualized microscopically in live cells. In budding

Two from One: A Short Introduction to Cell Division Mechanisms,
First Edition. Michael Polymenis.
© 2023 John Wiley & Sons Ltd. Published 2023 by John Wiley & Sons Ltd.

yeast, the onset of DNA replication correlates with bud formation, allowing one to tell whether a cell is in G1 (unbudded), or not (budded). To compensate for the lack of morphological landmarks associated with the different cell cycle phases, one can engineer them into cells. This is usually done by tagging proteins of established temporal abundance in the cell cycle with fluorescent tags. The presence or not of one or more fluorescence signals are then used as proxies for cell cycle transitions and the duration of cell cycle phases (Figure 3.1). The approach has been used not only in single-cell organisms, such as yeast (28) but in multicellular organisms as well. For example, fluorescent ubiquitinated cell cycle indicator (FUCCI) markers have been engineered into a variety of organisms (40–42), enabling real-time determination of

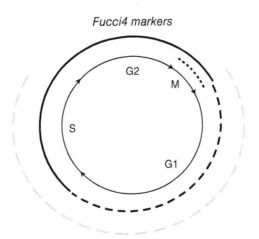

Figure 3.1 In the Fucci4 marker set, a combination of reporter proteins has been engineered to fluoresce with different colors in a way that can identify all four major phases of the cell cycle, based on the colors that appear or not. The different lines indicate the different markers and when they fluoresce in the cell cycle.

Source: Adapted from Bajar et al. (43).

the cell cycle position of individual cells in an animal and in culture. With single-cell information at hand, aggregate descriptions for the population can also be determined.

3.1.2 Labeled Mitoses

If fluorescent markers are not available, how can one measure the length of cell cycle phases from asynchronous cell populations? One way is to combine the visual landmark of mitosis with a molecular one for S phase. This is the basis of the old "labeled mitoses" technique (44, 45). Synthesis of DNA allows for the specific incorporation of a molecular marker in cells. Asynchronously dividing cells, in culture or even in animals, are given a pulse of tritiated thymidine. Only when they make DNA in the S phase will the cells incorporate the radioactive tritiated thymidine into their genome. So the pulse of tritiated thymidine will label a subset of cells, those in the S phase. The cells are then sampled at regular intervals afterward as they pass through mitosis, which serves as a fixed reference point, a milepost. Imagine that you only pay attention to cells in mitosis, and as you are watching them, you write down whether they are tritiated or not. What is measured is the fraction of cells in mitosis (by microscopy) that are also tritiated (after fixing the cells and quantifying the incorporated radioactivity in the DNA) (Figure 3.2).

The highest fraction of radioactive cells in mitosis will be observed when the cells that were in the S phase when the pulse was given progressed into mitosis. Mitosis is usually easier to score at the metaphase midpoint when the chromosomes are dense and aligned in preparation for anaphase. By the time they are scored as mitotic, these cells are already halfway through mitosis. Hence, the time from the end of the pulse until half of the labeled mitoses would be equal to the G2 phase's length plus about half the M phase length. The duration of mitosis

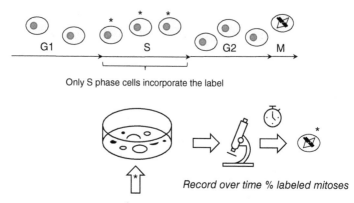

Only S phase cells incorporate the label

Record over time % labeled mitoses

Short pulse of ^{3}H-thymidine

Figure 3.2 Overview of the labeled mitoses technique.

can be measured microscopically. In human cells, the length of the M phase is fairly constant at about 30 minutes. The width of the labeled mitoses peak defines the length of the S phase. After the peak, there will be a valley because cells that were in G1 when the pulse was given will not be labeled, and their mitoses would be unlabeled. The second peak of labeled mitoses will appear when the labeled cells are in a second cycle (Figure 3.3). With this approach, the length of all the cell cycle phases can be estimated, including G1 (by subtracting the length of all the other phases from the total cell cycle time, shown as T in Figure 3.3).

The "labeled mitoses" approach assumes that the cells divide asynchronously, and the cycles are not unequal. It is also important that the concentration of the labeled thymidine added to cells is low, so that normal cell cycle kinetics are not affected. Nonetheless, the method is simple and highly informative. Its principle, scoring a marker for its presence through a downstream "milepost" or landmark of the cell cycle, is adaptable to many variations and detection methods.

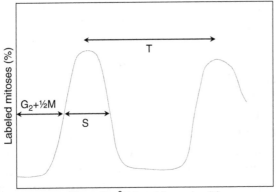

Figure 3.3 Expected graph of a labeled mitoses experiment. The length of each arrow indicated the length of the indicated cell cycle phase. T is the total cell cycle time.

3.1.3 Frequency Distributions

What if all one can do or is interested in doing is taking a snapshot of all cells and asking what fraction of cells are in a particular cell cycle phase? Imagine that you look at a slide of some fixed tissue and ask what fraction of cells are in mitosis. As it turns out, the information from such frequency distributions is valuable, and it can be obtained from any setting without the need for engineering cell cycle markers. For example, calculating the fraction of cells in mitosis (a.k.a. mitotic index) from cancer patient biopsies has diagnostic and prognostic value. Typically, the higher the mitotic index, the faster cells proliferate, and the worse the prognosis is. Another frequency distribution used in yeast is the budding index, the fraction of budded cells, and, therefore, not in the G1 phase. By far, however, the most common cell cycle frequency distribution reported in the literature is the amount of DNA per cell (DNA content). To measure how much DNA each cell has, the cells are first fixed and then stained by a fluorescent dye that binds DNA. A cell with fully replicated DNA will be in G2 or M phases, and it will fluoresce twice as much as a cell in G1, with unreplicated DNA. The remaining cells in the population will be in the S phase, and they will fluoresce with an intensity higher than the intensity of G1 cells but lower than that of G2 or M cells. Flow cytometry is the method of choice to measure the fluorescence per cell from thousands of cells accurately. As cells pass through a flow cell, they line up one at a time for sensing, which usually includes multiple lasers and fluorescence detectors. In addition to dye-specific fluorescence signals, the instrument collects information about cell size, based on how each cell scatters light.

A DNA content profile of an asynchronous cell culture is typically displayed as a histogram (see Figure 3.4). The result is frequency distributions of cells in G1, S, and G2 + M. The frequency distributions are measured from "area under the curve" calculations. These calculations are

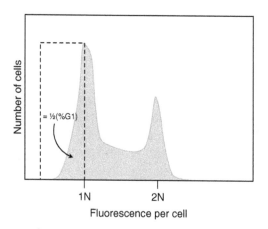

Figure 3.4 A hypothetical DNA content histogram, from haploid yeast cells. The cells under the 1N peak are in G1, and those under the 2N peak are in G2 + M. The area under the left half of the G1 peak can be multiplied by two, to estimate the percentage of cells in the G1 phase.

not always straightforward, especially when the G1 and G2 + M peaks are close to each other as in the example shown in Figure 3.4. Different models make different adjustments for the portion of the two peaks taken up by S phase cells. When accurate curve fitting becomes tenuous, a simple alternative is to measure only the left half of the G1 peak, which does not contain any S phase cells, and multiply it by two, to obtain the fraction of cells with G1 DNA content (Figure 3.4). If live cells in suspension have been engineered to express fluorescent markers in specific cell cycle phases, the frequency distributions can easily be obtained with flow cytometry. Furthermore, the fluorescence-activated cell sorting (FACS) capability of flow cytometers can identify and select cells in a particular cell cycle phase.

But just because a given fraction of cells is in a particular cell cycle phase does not necessarily mean that the *same* fraction of the cell cycle is taken up by that phase (46). That would only be true if the cell number remains constant, and for every cell that is generated, another one is removed. This situation happens in continuous microbial chemostat cultures. In other settings, however, we need to remember that there are always more cells early in the cell cycle in cultures that increase in numbers exponentially than in later phases.

Why that is, comes down to the fact that there will be two cells after cytokinesis from one cell closer and closer to cytokinesis. As a result, the fraction of cells in G1 is inflated a bit, while the fraction of cells in M is underrepresented (see Figure 3.5). For a cell culture that has a mitotic index of about 0.1, a properly weighted number from the graph shown in Figure 3.5 is about 0.14 (the calculations are given in (47)). Furthermore, unequal cell cycles, non-exponential growth and division, or cell death, all-cause deviations from idealized models. Despite all these caveats, a great deal of information is obtained

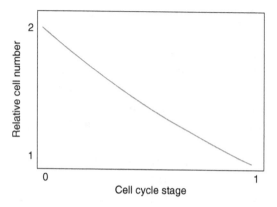

Figure 3.5 The relative cell numbers (*y*-axis) along the cell cycle (*x*-axis) from the beginning (point 0) to the end (point 1) of the cell cycle, follow the relationship $y = 2^{(1-x)}$.

from frequency distributions, even if estimates are rough and imprecise (46). For example, since the duration of mitosis in most somatic human cells does not vary much, any significant increases in the mitotic index in clinical specimens are usually attributed to faster cell proliferation associated with pathologies, such as cancer.

3.2 Growth Limitations and Variations in the Duration of Cell Cycle Phases

Cells growing in poor nutrients or with limiting growth factors will divide slower. Proliferating human cells in the body or cancer cells from different tissues divide at very different rates, with their cell cycle times varying from about 15 hours to more than 100 hours (48). In these situations, a question to ask is whether all the cell cycle phases proportionally increase in duration, or cells spend more time in a particular cell cycle phase. Important clues came from budding yeast. Remember that budding marks initiation of DNA replication in that organism. Counting the fraction of budded cells, the budding index, reflects the fraction of cells not in the G1 phase. Work in the Hartwell lab in the 1970s showed that cultures of the same cells in different nutrients have different budding indices. If all the cell cycle phases lengthened proportionally, the budding index should not change. Instead, in poor nutrients the budding index is lower, suggesting that cells spend disproportionately more time in the G1 phase (20). The same effect can be mimicked in rich nutrients if low doses of a protein synthesis inhibitor are added in the cultures. Years later, with the landmark discovery by Mike Hall and others of the TOR genes, the "master regulator" of multiple anabolic pathways including protein synthesis, it was clearly shown that upon loss of TOR activity cells arrest in early G1 with a small size. Likewise, removing serum and growth factors from the culture media of animal cells increases the fraction of cells in the G1 phase (49). Hence, if cell growth is inhibited, by removing nutrients or growth factors, or by chemical inhibitors, then cells delay initiating DNA replication. A significant portion of the variability in cell cycle times then arises from variations in the length of the G1 phase (Figure 3.6). Still, although the duration of the G1 phase is disproportionately affected by growth limitations, in budding yeast and many mammalian cells

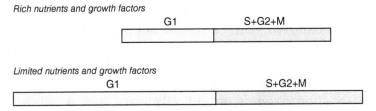

Figure 3.6 When cell growth is limited, the G1 phase of the cell cycle lengthens disproportionately.

non-G1 phases also lengthen, just not to the same extent as G1 does (48, 50). The length of the S and M phases does not change much, but the G2 phase can expand significantly when growth is limited in fission yeast and some animal cells.

The implications of these observations are significant. We have already seen that cell growth is required for cell division, and not the other way around. Since growth limitations disproportionately prolong some phase, then growth requirements for cell division may be disproportionately imposed in that phase (usually the G1 phase in budding yeast and animal cells, or G2 in fission yeast). Another way to see this is that cells monitor their growth status before a key cell cycle transition, say before initiation of DNA replication, perhaps by reaching a critical size threshold. If conditions are favorable (interpreted as being large enough), cells "commit" to the processes leading to their genome's duplication and segregation. In this view, we return to our discussion of the coupling between cell growth and division, about sizers and adders, and if and how cells monitor their size.

Commitment means there is no going back. As if cell growth triggers an irreversible switch, which propels cells forward to start and finish their division. Once cells pass through the commitment step in late G1, called Start in budding yeast or the Restriction Point in animal cells, they will complete the rest of the cell cycle transitions to cytokinesis, even in the face of nutrient limitations or growth factor removal. Quiescent cells in animal tissues or starved yeast cells have not passed through the Restriction Point or Start. Instead, they are thought to be in a non-proliferative G1-like state, called G0. The molecular mechanisms underpinning the commitment step and its relation to some metric of cell growth, such as cell size, are still unclear, actively debated, and will be discussed again later.

3.3 Synchronous Cultures

To study something that "cycles" and repeats itself, one has to distinguish the periodic oscillations in processes and molecules associated with cell division. Tracking individual cells can provide valuable information, but a single cell seldom provides enough material for downstream analytical assays. For many experiments, it is necessary to have enough cells that progress synchronously in the cell cycle.

In some systems, cell divisions are naturally synchronous. The rapid, early embryonic cell cycles of animals are synchronous and proceed without cell growth. Such systems of natural synchrony are an excellent biological resource. It is no coincidence that molecules that govern cell cycle transitions (e.g., cyclin proteins) were first identified in those settings (51). The abundant material and natural synchrony of egg extracts from *Xenopus laevis* frogs, sea urchins, and other organisms have been instrumental and influential tools in cell cycle research. They will continue to be so. In most other cases, however, populations of cells in culture or tissues are randomly distributed at different cell cycle phases. In principle, there are two general ways to get synchronized cells from asynchronous cultures: Either induce all cells to divide synchronously or select only a small fraction of cells found exclusively in a particular cell cycle stage.

3.3.1 How can One Induce Synchrony?

Imposing some block in the cell cycle will cause all cells in an asynchronously dividing culture to accumulate and arrest at a specific point in the cell cycle. Then the cells are released from the block and allowed to progress synchronously in the cell cycle. But while the cells are arrested, they continue to grow in mass. Their growth is "unbalanced" from their division. This is a significant disadvantage, which, if not adequately controlled, could lead to artifacts. A wise strategy is to follow two synchronous cell cycles after cells are released from the block. The advantages of arrest-and-release methods to induce synchrony are that they are simple, and the yield is high. Potentially all the cells in a culture can be induced to cycle synchronously. In practice, some cells do not recover from the arrest, and this is another variable that needs to be taken into account.

There are several ways to arrest the cell cycle. If temperature-sensitive mutants that arrest at a specific point in the cell cycle exist, then all one has to do is shift the culture to the non-permissive temperature. After about one or two cell cycle times, all the cells in the culture would be arrested at the same point in the cell cycle. When the cells are shifted back to their permissive temperature, the cells that recover from the arrest will progress synchronously from that point on in the cell cycle. Such mutants, the *cdc* collection, were first discovered by Lee Hartwell in budding yeast (52), and they now exist in other organisms, but not in most animal systems.

Chemicals that impose a specific cell cycle block are used widely. Such inhibitors' ideal attributes as synchronization agents are specificity, low toxicity, and full reversibility of the block after washing the cells. Compounds that destabilize microtubules (e.g., nocodazole, colcemid), interfere with spindle formation, and cells cannot complete mitosis. Aphidicolin is an inhibitor of DNA polymerase, leading to S phase arrest. Inhibiting DNA replication is also relatively easy to achieve by perturbing the nucleotide pools needed to make new DNA. For example, hydroxyurea inhibits ribonucleotide reductase, which lowers the deoxyribonucleotide triphosphate (dNTP) levels in the cell. A common approach to synchronize mammalian cells through inhibition of DNA replication is to expose them to high concentrations (e.g., 2 mM) of thymidine. Excess thymidine alters the dNTP pools, with a marked decrease

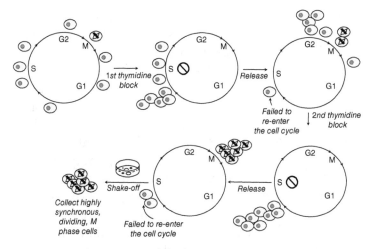

Figure 3.7 Releasing the cells from a double thymidine block is a popular approach to induce synchrony in animal cells. When combined with mitotic shake-off, highly synchronous dividing cells can be selected.

of dCTP levels, probably through feedback inhibition of the metabolic pathway associated with dCTP synthesis. Adding excess thymidine to an asynchronous culture will arrest them at the point in the S phase they happen to be. Cells not in the S phase when thymidine was added will progress until they reach the next S phase and arrest at the G1/S boundary (Figure 3.7). After a single thymidine treatment, all the cells will be arrested in the S phase, but not uniformly at the same point of the S phase. Washing off the thymidine will release the cells from a single thymidine block, and in a few hours, all the cells will complete and exit the S phase. After they exit the S phase and before having enough time to reach the next S phase, treating them again with a second dose of thymidine will arrest them in late G1 just before the onset of the next S phase. Hence, a single thymidine block will cause a broad S phase block, but a double thymidine block will lead to a narrow late G1 block (Figure 3.7).

Other cell cycle inhibitors include compounds that inhibit the central enzyme that drives cell division, the cyclin-dependent protein kinase (Cdk), which we will describe in detail in the next chapter. A popular such inhibitor, RO-3306, is a highly selective and reversible one, which arrests cells at the G2/M transition (53). Washing off the compound enables the vast majority of the cells (~95%) to synchronously enter the M phase. In yeast, a standard method to synchronize haploid cells is to expose them to alpha-factor, a mating pheromone produced by cells of the opposite mating type. Pheromone triggers a signaling pathway with outputs necessary for mating, including inhibition of Cdk at the G1/S transition (successful mating is not compatible with ongoing genome duplication and segregation).

The above list of inhibitors of cell division is not exhaustive. Additional compounds exist, and more will be discovered in the future. However, they all have in common that during the cell cycle block, cell growth continues. A different approach is to exploit this relation between cell growth and division. As we have already seen, removing nutrients or growth factors will disproportionately delay cells in the G1 phase. Left for enough time under such limiting conditions, most cells will eventually be in a quiescent state in early G1. Most of these cells will progress into late G1 and S phases when they are re-fed with nutrients and growth factors. However, cellular physiology is profoundly affected by the nutrient and growth factor changes. The synchrony is also lower compared to inhibitor blocks. Some cells do not re-enter the cell cycle, and those who do may progress through cell cycle transitions at different rates.

3.3.2 Selecting for Synchrony

Imagine that from a culture with cells distributed at random all over the cell cycle, you can magically reach in and grab only the ones at one specific point. When you start a culture with just these cells, they will continue progressing synchronously in the cell cycle. As long as you did not damage or stress them during the selection process, their growth and division will be balanced, and cellular physiology is maintained. These are critical advantages, but they come at a steep price. You end up with a much smaller cell population than the population in the original culture.

Selection methods separate cells based on some cell cycledependent changes in their physical properties. A typical selection method is mitotic shake-off when working with adherent cells requiring a solid surface to grow in culture. Cells in mitosis become rounded and less adherent to the surface of the solid support of a Petri dish. All one has to do is gently shake the dish and collect the supernatant, which only contains cells in the M phase. The simplicity of the method, the exceptional synchrony, and cellular physiology maintenance are significant advantages. The low yield (only a small fraction of cells are in mitosis at any given time), and the requirement of working with adherent cells are its limitations. Nonetheless, the approach can be used in combination with other blocks, such as after cells are released from a double thymidine block, increasing the yield of highly synchronous, cycling cells (Figure 3.7). Lastly, with the recent development of fluorescent cell cycle reporters (e.g., see the earlier discussion of the FUCCI system; Figure 3.1) one could also apply FACS to isolate cells in a particular phase. However, the longer the phase chosen, the poorer the synchrony. Since the M phase is the shortest phase, isolating cells in M phase will yield the most synchronous populations.

3.3.2.1 Elutriation: The Mother of all Synchrony Selections

The most powerful cell separation method used to select cells for synchronous cell cycle studies is centrifugal elutriation. The technique works only with cells that proliferate

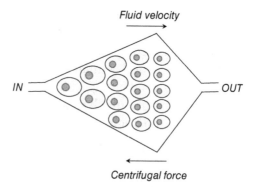

Figure 3.8 Centrifugal elutriation separates cells in a continuous flow centrifuge mostly based on their size.

in suspension. It relies on the sedimentation properties of cells, which are directly related to cell size. The smallest cells in a culture will usually be newborn daughter cells. Elutriation has to do with sedimentation, but the cells are not pelleted. The asynchronous cell suspension is loaded into the rotor chamber of a continuous flow centrifuge. As cells are centrifuged, media is also pumped in. If the pump rate and centrifugation speed are chosen correctly, the cells will come to rest in solution in the chamber (Figure 3.8). This happens because the outward inertial forces from the centrifugation are balanced by the inward flow and buoyancy from the media's continuous flow. As a result, the cells are kept in suspension and distributed across the chamber based on their physical properties and the velocity gradient in the chamber. Numerous variables affect the cells' separation in the chamber (e.g., rotor speed, chamber shape, counterflow, fluid velocity in the rotor chamber, cell load, density, and viscosity of the elutriation medium). Still, all other things being equal, the cells separate based on their size. By carefully adjusting the centrifuge rotor's speed and the media's flow rate in the chamber, the cells are selectively coming out of the chamber and collected. Although the cells probably experience some stress during the centrifugation, they are never arrested and continue to progress in the cell cycle during the entire experiment. Nonetheless, elutriation requires specialized instrumentation, and it is a selection method. Hence, elutriation also suffers from the low yield of synchronous cells.

Since centrifugal elutriation separates cells based on size, we can use the relationship between cell size and cell cycle position to measure the length of the G1 phase. If cells actively monitor their size and do not initiate DNA replication until they reach a critical size threshold, cell size can be used as a proxy for cell cycle position. This is obviously a sizer situation, which, as we described earlier, is the case for fission yeast and daughter cells of budding yeast. In systems with a sizer, cells of the same size ought to be at the same point in the cell cycle. Note that the cell size in elutriation experiments is always measured at regular intervals with a Coulter counter. If we know the size at which budding yeast cells are born (V_b), the size at which they initiate DNA replication (V_d; scored visually by the appearance of buds), and the rate at which they increase in size (k; calculated from the plot of cell size as a function of time), then the length of the G1 phase (T_{G1}) can be obtained from the following equation: $\ln(V_d/V_b) = kT_{G1}$ (Figure 3.9).

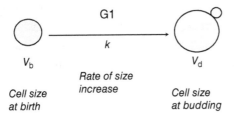

G1

k

V_b

Rate of size
increase

V_d

Cell size
at birth

Cell size
at budding

Figure 3.9 In budding yeast daughter cells, which increase in size exponentially, we can estimate the absolute length of the G1 phase from their size at birth (V_b), their size at budding (V_d; marking the end of the G1 phase), and how fast they increase in size over time (k).

For systems with a sizer behavior, it is possible to solve the low yield of synchronous cells from a single elutriation. To increase the number of cells at a particular cell cycle point, all then one has to do is to do multiple elutriations. Individual elutriated samples are allowed to progress in the cell cycle and harvested at the desired cell size. The harvested samples, which originated from different elutriations, can be combined into pools, as long as these pools comprise populations with the same average cell size. All the cells in each pool will be at the same point in the cell cycle. A cell size series covering the entire cell cycle can be generated in this way, with enough cells at each point for downstream analytical assays (54–56). The approach is laborious, but the cell size series is indistinguishable from analogous time series from the same cells and medium. However, it can only be properly implemented in systems with a sizer.

Overall, it is hopefully becoming apparent that our earlier discussion about the coupling of cell growth with division has implications for numerous aspects of cell cycle studies. For example, we saw how in induction, arrest-and-release methods to synchronize cells, cell growth is not balanced with cell division. On the other hand, in selection methods, such as elutriation, the strong coupling between growth and division generates valuable landmarks/mileposts that allow us to know where cells are along their cell cycle. In the next chapter, we will see how the coupling between growth and division also led to discovering key parts of the cell cycle enzymatic engine.

4 The Master Switch

- How was Cdk discovered?
- Is the regulator the same in all cells?
- How were cyclins discovered?
- Cell cycle transitions in cell-free systems

In this chapter, we will see how the two central subunits of the cell cycle switch (Cdk and cyclin) were discovered. Cdk is a catalytic subunit, a protein kinase, that by itself is inactive, unless it binds to an activating regulatory subunit, a cyclin. It can be challenging to keep track of the names of key mutants and gene products in the various organisms. There is a table organizing most of the names of the gene products mentioned in this chapter (see Table 4.1). A larger list, with all the proteins and their function mentioned in all the chapters can be found at the end, before the References (see Table 9.1).

Two from One: A Short Introduction to Cell Division Mechanisms,
First Edition. Michael Polymenis.
© 2023 John Wiley & Sons Ltd. Published 2023 by John Wiley & Sons Ltd.

Table 4.1 Mutants, proteins, and associated information, as mentioned in this chapter.

Mutant	Phenotype	Protein	Protein function	System
cdc28-1	Loss-of function (large size, unbudded, G1 arrest)	Cdc28	Cdk (cyclin-dependent kinase)	Budding yeast
cdc28-1N	Loss-of-function (large size, budded, G2 arrest)	Cdc28	Cdk	Budding yeast
wee1	Loss-of-function (small size)	Wee1	Kinase that inhibits Cdk	Fission yeast
wee2	Gain-of-function (small size)	Cdc2	Cdk	Fission yeast
cdc2	Loss-of-function (large size, G2 arrest)	Cdc2	Cdk	Fission yeast
cdc25	Loss-of-function (large size, G2 arrest)	Cdc25	Phospha-tase that activates Cdk	Fission yeast
CLN3-1	Gain-of-function (small size)	Cln3	G1 cyclin	Budding yeast
cln3Δ	Loss-of function (large size)	Cln3	G1 cyclin	Budding yeast
cln1,2,3Δ	Loss-of function (large size, G1 arrest)	Cln1,2,3	G1 cyclins	Budding yeast

Table 4.1 (Continued)

Mutant	Phenotype	Protein	Protein function	System
		Clb1-6	Mitotic (B-type) cyclins	Budding yeast
		MPF	Complex between Cdk and mitotic cyclin	Animal embryonic cell cycles

4.1 Genetic Analyses Leading the Way

4.1.1 The *cdc28* Mutant of Budding Yeast

The isolation of the cell division cycle (*cdc*)[1] mutants by Lee Hartwell and his students 50 years ago was a watershed moment in cell cycle studies and, more generally, genetics. It epitomizes the beauty and "awesome power of yeast genetics," where so much can be learned from simple means and clear ideas. As always, the choice of model system matters. It has to be suited to the problem. Hartwell chose *Saccharomyces cerevisiae* for several reasons:

1. *General* reasons that have drawn others to yeast before and after him: It is a microbe, genetically tractable, easy, and cheap to work with.

---

[1] A note on nomenclature. *S. cerevisiae* gene names are italicized, and when referring to loss-of-function mutants they are in lower case. The protein gene products are not italicized, and the first letter is capitalized. A gene deletion is indicated with the Greek letter Δ.

2. A *problem-specific* reason distinguishing budding yeast from other competing model systems: The bud (present or not and its size) provides the all too valuable cell cycle landmarks and mileposts one needs to track the process under study (see Figure 4.1). With a simple light microscope, Hartwell could monitor cell cycle progression.

3. The largely *aspirational* reason at the time was that since yeast is a eukaryote and cell division is so fundamental, the machinery that drives the cell cycle may be similar to that of other organisms, including humans.

Armed with a microscope and a collection of temperature-sensitive yeast mutants, which he made following classical mutagenesis approaches, Hartwell then had a clear idea that opened up the field. All his mutants would arrest in the cell cycle when shifted from room temperature to 37 °C. But a lot of things can kill an organism. There is death from the inability to perform a specific vital task you are looking for (in this case, completing a cell cycle transition) and death from any other cause. How do you separate the specific cell cycle mutants from all the others? This is where Hartwell's lab made the key realization and exploited the unique properties of the organism. Unlike cells from almost all other organisms, individual yeast cells look different depending on where they are in the cell cycle. Unbudded in G1, with tiny buds as they initiate DNA replication, dumbbell-like in mitosis, and so on. Focusing on *how* they looked under the microscope when arrested, Hartwell could distinguish mutants that could not perform a specific cell cycle task from mutants that could not perform any other function essential for life (e.g., central metabolism, etc.). Mutants that could not complete one specific cell cycle point would keep progressing in the cell cycle until they reach that point, and then all cells will have the same, uniform arrest morphology (Figure 4.1, right). In contrast, the cells of mutants that could not perform a function needed at multiple cell cycle points would not progress much in the cell cycle when shifted to 37 °C. They will be arrested before each of these points, leading to a nonuniform arrest of the mutant population as a whole (Figure 4.1, middle).

To illustrate with an analogy, imagine that you want to know the vital parts of your car's engine and how they work, and imagine scenarios where your car would stop running (i.e., "mutant phenotypes"). From which of these scenarios would you gain more information about specific engine parts? Running out of gas, or from a defective gearbox? Hartwell realized that he could differentiate general essential functions (cells "running out of gas," etc.) from more specific ones (cells switching to the next cell cycle stage, "gear"). The screen had a simple binary outcome (Figure 4.1): Cells arrest with a *random* budding morphology (mutants in bulk essential functions needed at all or most points of the cell cycle), or they arrest with

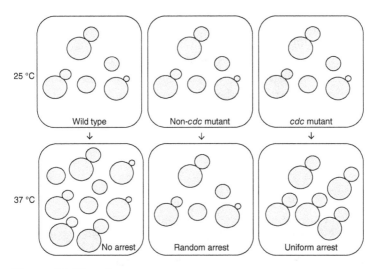

Figure 4.1 The *cdc* screen.

a *uniform* budding morphology (mutants in an essential function needed at one point in the cell cycle) (52).

Hartwell identified >30 different *cdc* mutants (there are now >100). We will mention several of them later. He classified them further based on other properties. Progression through the G1/S transition triggers initiation of DNA replication, budding, and spindle pole body (SPB) duplication. The yeast SPBs are analogous to the centrosomes of animal cells, forming the poles of the microtubule spindle that segregates the chromosomes in mitosis. Some mutants were impaired in one of these processes when shifted to 37 °C. For example, in budding, in DNA replication, or in SPB duplication. But *cdc28* mutants were impaired in all three. When shifted to 37 °C, *cdc28-1* cells arrest without a bud, with unreplicated DNA, with one SPB, and at the same point that the mating pheromone arrests the cell cycle (Figure 4.2). Hence, similarly to mating pheromone arrested cells, *cdc28-1* cells were also arrested as large unbudded cells, arguing that cell growth was unaffected, even though major cell division-related processes at the G1/S transition were blocked.

While other gene products ruled one cell cycle function, Cdc28 appeared to rule them all, without significantly inhibiting cell growth. Could it be that Cdc28 was part of a master controller of cell division? It would take many years to reach that conclusion, from work in many systems.

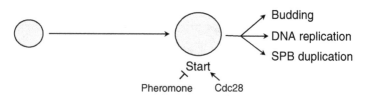

Figure 4.2 Budding yeast cells commit to a new round of cell division in late G1, at Start. After they complete START, at the G1/S transition they will bud, duplicate their SPBs and initiate DNA replication.

Nonetheless, Hartwell proposed that Cdc28 controls the commitment step, Start, in late G1 (20, 52), and dictates the onset of cell cycle-dependent processes. Cells that have completed Start are refractive to mating pheromone. They go on and complete cell division, and they will be arrested for mating at the next cell cycle. All these phenotypic observations suggest that Cdc28 must be at the receiving end of whatever pathways allow cells to monitor their growth status and commit to division. Since its discovery in the early 1970s, the *cdc28* mutant seemed special. By the early 1980s, the gene was cloned (57), and shown to encode a protein kinase that phosphorylates serine and threonine residues (58).

4.1.2 From the *wee1* to the *cdc2* Mutant of Fission Yeast

The Hartwell screen was transformative because it showed that the cell cycle could be studied by powerful, genetic means. The same approach could be used in other systems, provided that some morphological milepost of cell cycle progression is available so that the mutants can be properly interpreted. But as we have seen, other than mitosis, there isn't much else to use in other organisms. Except, of course, exploiting the relationship between cell growth and division and using cell size as a metric. For that relationship to be of value in a cell cycle genetic screen, two criteria must be met: First, cell size must closely reflect the cell cycle position. In other words, the cells of that organism must display a strong sizer behavior and actively sense their size before they complete a cell cycle transition. Second, cell size must be easily measured, preferably by a quick visual inspection. On both counts, the fission yeast *S. pombe* is ideal (see Figure 2.7). As we have already discussed, fission yeast has the most robust sizer behavior among any system. Furthermore, because the cells are cylindrical and elongate

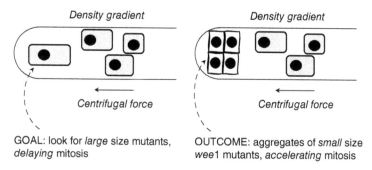

GOAL: look for *large* size mutants, *delaying* mitosis

OUTCOME: aggregates of *small* size *wee1* mutants, *accelerating* mitosis

Figure 4.3 The unexpected recovery of the *wee1* mutant.

as they grow without changing their diameter, cell size can be literally measured with a ruler.

In the 1970s, shortly after Hartwell's lab published their results, Paul Nurse set out to find cell cycle mutants in fission yeast. He reasoned that a specific block in some cell cycle process would not block cell growth, and the mutants, when shifted from room temperature to 35 °C, would be large. He could then isolate these mutants physically through a density gradient (Figure 4.3). Single large cells would be at the bottom of such a gradient, while small ones would be at the top. But when he looked at one of these mutants at the bottom of the gradient, not only were the cells not larger, they were smaller. A lot smaller. They were also not arrested in the cell cycle. Instead, they divided at half the standard size. Their small size phenotype was temperature-enhanced, not temperature-sensitive. The title of that paper, "Genetic control of cell size at cell division in yeast" summed it all up (59). But how did that mutant, *wee1*, end up at the bottom of the gradient in the first place? Because the mutant cells had a secondary defect and aggregated at 35 °C. As Louis Pasteur put it, "...chance favors only the prepared mind." Nurse recognized what he had in his hand and used it to dig deeper into cell cycle control mechanisms. The small size (*wee*) mutant phenotype suggested that specific genetically-controlled steps in the cell cycle might be rate-limiting for cell cycle progression.

More selections for small-size mutants yielded many more mutant alleles of *wee1*, plus just one more mutant that mapped at a different locus, and it was named *wee2*. The mutant alleles, *wee1* and *wee2*, also behaved differently in the presence of their corresponding wild-type alleles, in heterozygous diploid cells. Whereas *wee1* was recessive, *wee2* was dominant. These functional relationships suggested that the *wee1* mutants were loss-of-function ones, but *wee2* was a gain-of-function mutant. Hence, in wild-type cells, the *wee1*$^+$ gene product is not required for mitosis but instead inhibits it. That explains why when its function was lost in the many *wee1* mutants Nurse identified, cells entered and completed mitosis. They even did so at a smaller size. On the other hand, based on the single dominant *wee2* allele's behavior, in wild-type cells, *wee2*$^+$ would likely be a positive regulator of mitosis.

Keep in mind that while a cell cycle transition accelerated in *wee* mutants, in *cdc* mutants, cell cycle transitions are blocked. Do different phenotypes (acceleration vs. block) always arise from entirely different genes? The answer was no. It was surprising and exhilarating to find that *wee2* mapped to the *cdc2*$^+$ locus in fission yeast (60). The rare, dominant, gain-of-function *wee2* mutation promotes mitosis. The more common, loss-of-function, recessive mutations in the same gene arrest cells at mitosis, as would be expected for *cdc* mutations. Different mutations in the same gene could either prevent or accelerate division, suggesting that the Cdc2 protein is a critical regulator of mitosis (61). It is required for cell division, *and* it can also accelerate it. What else would you want from a master regulator? Maybe that it is also universal?

4.1.3 What is True for One is True for All

Finding a crucial part of one engine is one thing. But finding the same part in all engines is quite another. Then maybe all such engines run the same way. From the mid to late 1970s, cloning genes became possible. What ensued was the "great cloning orgy," which lasted until the mid-1990s. During that time, the genes of most cell cycle mutants were cloned. But the way they were cloned in some cases was telling. Not only budding yeast *CDC28* and fission yeast *cdc2+* encode enzymes (protein kinases) that look similar on paper but you can substitute the function of one for the other in vivo. The experiment is simple. All you have to do is ask if there are any DNA pieces from budding yeast that could rescue arrested fission yeast *cdc2* mutant cells at 37 °C (Figure 4.4). The answer was yes, and the piece of DNA had the *CDC28* gene on it (62)! If such functional complementation works between yeasts

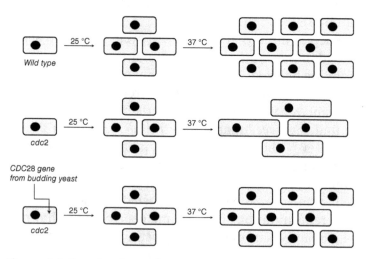

Figure 4.4 Functional complementation and conservation of the cell cycle engine. Genes from one organism can substitute for the missing cell cycle function in another, distantly related organism.

that diverged 300–600 million years ago, why stop there? Could it also work between yeast and human cells? Sure enough, that was the case (63). That genetic screens in yeasts would lead to conserved parts of the cell cycle engine seemed like a pipe dream in the early 1970s. In the 1980s it was a reality, with the human orthologues of Cdc28 and Cdc2 at hand.

4.2 All Roads Lead to the Same Control System

So far, we have looked at genetic approaches that led to the master controller of cell division. But the cell cycle has always been propelled by a combination of methods and systems. In the 1980s, all these different approaches were aligned for a spectacular convergence.

4.2.1 Cyclins

Much of the experimental work on the cell cycle comes down to understanding *what* molecules, activities, structures, etc. go up and down in the cell cycle, *how*, and *why*. The discovery of *cdc* genes was a major milestone, but it did not change the fact that few macromolecules were known to vary in abundance in the cell cycle other than DNA and histones. It's all about landmarks and mileposts again, this time, molecular ones. The discovery of cyclin proteins is yet another example of chance observations paving the way to fundamental insights. The experimental system here was the early embryonic cell cycles of animals. The large unfertilized oocyte is arrested at the metaphase of the second meiotic division. It will stay arrested unless it is fertilized by sperm. When it is penetrated by sperm, it will complete meiosis II and become a zygote (Figure 4.5).

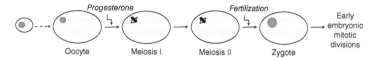

Figure 4.5 Oocyte maturation.

There is no new RNA synthesis during the rapid, synchronous, early embryonic cell cycles that follow. The maternal mRNAs previously stored in the oocyte instruct new protein synthesis. All gene expression control in that system is posttranscriptional, and fertilization activates protein synthesis. All that makes it a great system to study translational control. And it was protein synthesis that Tim Hunt wanted to study, using sea urchin eggs, which can be easily and cheaply collected. A simple experiment Hunt was doing was to give the eggs a short pulse of radioactive methionine, which would be incorporated into newly made proteins, and then separate those proteins on a gel. He wanted to see if the pattern after fertilization with sperm was the same as after some parthenogenetic treatments that also stimulate protein synthesis. He could then identify proteins whose synthesis might be important in the embryogenesis of invertebrates. Remember that the experiment was done over a period when the zygote undergoes rapid, synchronous divisions. What would you expect to see on those gels?

Based on what we have already discussed about growth patterns in the cell cycle, the intensity of the protein bands should get stronger and stronger over time. But one band's intensity did not. It came up and then went away about 10 minutes before cell division (51). It did so not just once but before every cell division (Figure 4.6). Hunt also showed that proteins with similar behavior, appearing and disappearing in a cell cycle-dependent pattern, existed in distantly related organisms. Except that in clam eggs, a couple of bands oscillated, not just one as in sea

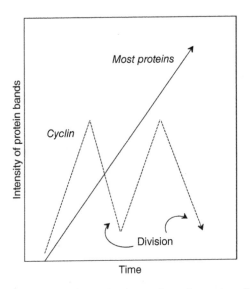

Figure 4.6 Cyclin abundance in the early embryonic cell cycles. After egg fertilization, the abundance of synthesized proteins was evaluated by radioactive methionine incorporation, followed by gel electrophoresis and autoradiography. The intensity of most proteins (solid line) showed no cell cycle changes and increased constantly. The intensity of cyclin, however, was periodic, rising and then falling, in the cell cycle (dashed line). The time when most cells were undergoing division in this experiment is indicated.

Source: Adapted from Evans et al. (51).

urchins. The proteins were aptly named "cyclins," and such dramatic molecular landmarks of the cell cycle were now a reality. Cyclins were made constantly through these cell cycles (keep in mind that there is no new RNA synthesis in early embryonic cell cycles and no G1). Their periodic pattern came down to their destruction before each cell division. At that point, there was no evidence that cyclins were causing anything to happen in the cell cycle. At a minimum, however, they were valuable tools. Molecular signatures and mileposts that one could follow in the cell cycle.

Why were cyclins not seen previously? Hunt was interested in protein synthesis in general. The cyclins' periodic disappearance act would not have been detected in asynchronously dividing cells, so the choice of system with its natural synchrony was a critical but serendipitous feature. But working with these eggs was not new, and Hunt did not use any cutting-edge technology. Protein separation on acrylamide gels was around for years. Maybe others before Hunt focused on the vast majority of proteins whose intensity did not oscillate. If your frame of mind is to match proteins that increase in intensity with the overall activation of protein synthesis, then most people probably did not think outside the proverbial "box." Thankfully, Hunt did.

4.2.2 Maturation Promoting Factor (MPF)

Another approach used to study the cell cycle was to ask what would happen to cells at some cell cycle point if they were injected with extracts from the same cells at a different cell cycle point (Figure 4.7). The goal is to identify factors that could drive or block cell cycle transitions. Over the years, the logic of this experiment was applied in several forms, from fusing cells at different cell cycle stages to injecting extracts into live cells, and,

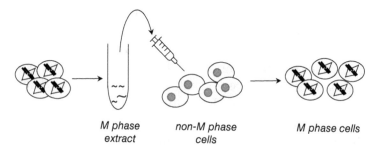

M phase   non-M phase   M phase cells
extract   cells

Figure 4.7 Mitosis is dominant over all other phases of the cell cycle. Injecting extracts from M phase cells into cells at any of the other phases of the cell cycle will trigger mitosis in the recipients.

finally, going entirely cell-free. The current cell-free sys-tems arose from such approaches. Frog egg extracts can recapitulate cell cycle transitions (e.g., forming a nuclear envelope, chromatin decondensation, initiation of DNA synthesis, and chromosome condensation) (64). They enable a biochemical attack on the problem. Much as the discovery of cell-free yeast extracts capable of glycolysis and fermentation at the turn of the twentieth century, at the dawn of biochemistry, or Avery's experiment in the 1940s showing that DNA is the genetic material.

From cell fusion experiments, it became clear that mitosis is dominant over all other cell cycle phases. Fus-ing cells in G1, S, or G2 phases with cells in the M phase will always induce mitotic landmarks (e.g., chromosome condensation) (65). Similar conclusions were reached by following egg maturation. Usually, premature oocytes enter meiosis I when stimulated by hormones, such as progesterone. The mature oocytes will then arrest at metaphase of meiosis II. If they are fertilized, they will complete meiosis II, form the zygote and start dividing mitotically. The "maturation" of oocytes is another way to say that they entered meiosis. Maturation can be trig-gered without progesterone, if cytoplasm from meiotic, "mature" oocytes are injected to premeiotic oocytes (66). Hence, the meiotic state of mature oocytes is dominant. Mature oocytes must contain a "maturation promoting factor" (MPF) that can drive oocytes to become mature, unfertilized eggs. Progesterone can trigger MPF in prema-ture oocytes, even if protein synthesis is inhibited, arguing that MPF components are already present and activated by some posttranslational pathway triggered by progesterone. It turns out that several cell sources could be MPF donors and drive the maturation of oocytes to unfertilized eggs. First, eggs without a nucleus could still produce MPF, arguing that MPF activation is an entirely cytoplasmic process. It is something in the cytoplasm that drives the different cell cycle stages of the nucleus. Second, MPF was not limited to oocytes. Cytoplasm from somatic cells arrested in the M phase could also drive oocyte maturation or meiosis. Hence, MPF is an activity present in most, if not all, mitotic cells. The "M" in MPF now stands for maturation, meiosis, and mitosis promoting factor.

Scientists now had an activity, MPF, that could be followed during synchronous cell cycle transitions. When Hunt was doing his cyclin experiments, Marc Kirschner's lab already had preliminary data indicating oscillations in frog MPF activity during oocyte maturation. MPF activity disappears rapidly at the end of the first meiotic division. In the second meiotic cycle, MPF reappears before metaphase, when the oocytes reach the mature unfertilized egg stage. We already saw that protein synthesis was not required to activate MPF when the oocytes were induced by progesterone to enter meiosis I. But MPF never reappeared in the second meiotic cycle if protein synthesis was inhibited. The simplest explanation was that premade MPF components are activated by progesterone and then destroyed as cells exit the first meiotic division. They must be synthesized anew for cells to reach the metaphase of the second meiotic division.

Furthermore, MPF activity dropped precipitously at fertilization, but then it returned in the subsequent rapid embryonic cell cycles. MPF activity rose and then disappeared before each cell division (67, 68). The pattern of MPF's appearance and disappearance matched that of cyclin's (Figure 4.8). Could it be that it is cyclin that must be made before mitosis so that MPF can be activated? Indeed, that was the case. Once cyclins were cloned, injecting the cyclin mRNA into cells arrested at G2/M drove them into M phase (69, 70). Why stop there? With cell-free extracts from frog eggs capable of undergoing multiple cell cycles in a test tube, things could be removed and added at will. Destroying cyclin mRNA arrests these extracts, arguing

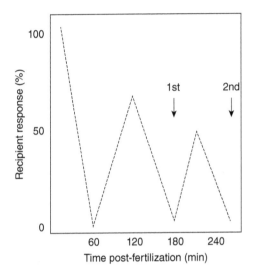

Figure 4.8 MPF oscillations. On the *x*-axis is the time after fertilization when the egg cytoplasm was assayed for MPF activity in recipient cells. On the *y*-axis is the percentage of recipient cells undergoing nuclear envelope breakdown, marking mitosis. Arrows indicate the time of cleavage divisions.

Source: Adapted from Wasserman and Smith (67).

that cyclin synthesis is necessary to activate MPF (71). In some legendary experiments done by Andrew Murray in Kirschner's lab, they asked if cyclin synthesis is all it takes to activate MPF (Figure 4.9). First, they destroyed endogenous mRNAs. Careful RNase treatment degrades mRNAs but not the well-folded tRNAs and rRNAs involved in protein synthesis. Then, adding back just cyclin mRNA alone was enough to drive multiple cell cycles.

The newly synthesized cyclin protein accumulated during each interphase and was degraded at the end of each mitosis (71). Is it also cyclin that needs to be destroyed for cells to exit mitosis and complete cell division, inactivating MPF? Sure it is! Adding mutant cyclin mRNA, encoding a proteolysis-resistant cyclin protein that lacks the first 90

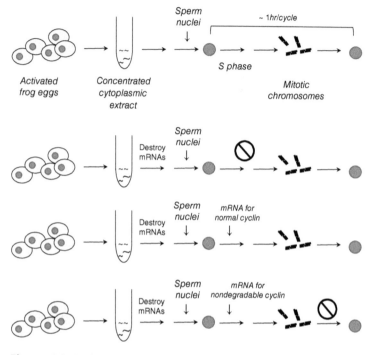

Figure 4.9 Cyclin is necessary and sufficient to trigger mitosis in frog oocyte extracts. But to exit mitosis, cyclin must be destroyed.

amino acids, drives the cell free-extracts to mitosis, but they stay there arrested and never return to interphase (72). That was the definitive evidence that cyclin synthesis turns MPF on, while cyclin degradation turns it off (Figure 4.10). Is the cell-free system powerful enough for you yet?

Could it also be that cyclin and MPF are parts of the same master switch and, if so, how are they related to Cdc28 and Cdc2? By the late 1980s, Cdc28/Cdc2 and cyclin orthologues seemed ubiquitous in eukaryotes. After many years and painstaking biochemical fractionation schemes from frog eggs, MPF was purified (73), and unity emerged. MPF was a dimer with protein kinase

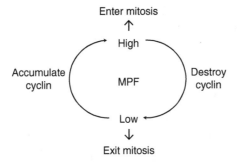

Figure 4.10 Cyclin synthesis turns MPF on, while cyclin degradation turns it off.

activity, containing the frog versions of Cdc28/Cdc2 (74) and cyclin (75). Without the cyclin, the kinase subunit has no activity. Cdc28, Cdc2, and their relatives in other organisms define the Cdk family. Once Cdk is bound to cyclin, its active site is opened up and properly configured for catalysis (76). The active enzymes are complexes of one catalytic (Cdk) and one regulatory (cyclin) subunit. All the previous 20–30 years' work reached a spectacular crescendo, offering one of the best examples in biology about the synergy among different approaches and model systems, which led to the grand prize of the master switch of cell division. The rest of the material we will cover will be related to the master switch, in one way or another. But before we close this section, we will return to some parts of the Cdk/cyclin narrative, which were left open.

4.3 Making Sense of it All

By now, you should feel awed by the beauty and thrill of the discoveries that led to the master switch. Then, as you start thinking about it a little more, you should also have some unanswered questions: Why did the original *cdc28* mutant in budding yeast arrest at Start? What is Wee1, and how did it inhibit Cdc2 in fission yeast? Answering these questions reveals more control layers (but not all) of the cyclin/Cdk switch.

4.3.1 Cyclins Galore in Budding Yeast

At first, there was confusion about why *cdc28* arrests in late G1, while its fission yeast counterpart, *cdc2*, arrests later in the cell cycle. We now know that these Cdks, in both organisms, regulate all cell cycle transitions. But there are also "lifestyle" differences between the organisms. As we have discussed, in rich nutrients, fission yeast has a robust G2 sizer, while a G1 sizer operates in budding yeast daughter cells. Most *cdc28* loss-of-function mutants arrest at Start, but some (e.g., *cdc28-1N*) do arrest in G2 (77). The *cdc28-1N* was also a valuable tool in genetic screens for regulators of the G2/M transition. A general strategy in genetics is to combine genetic alterations and see if and how the original mutant phenotype is modified. If the wild-type phenotype is restored, then you have identified suppressors. In this case, it was asked if there are gene products that, when given in high dosage, could rescue *cdc28-1N* cells from arrest at 37 °C (77). Remember that in the mutant, the Cdk is still there, albeit crippled. Among the hits were genes similar to the mitotic cyclins that Tim Hunt had discovered. Budding yeast has four of them (Clb1,2,3,4)! And that is not all. As we will see, budding yeast has five more cyclins, with distinct expression patterns in the cell cycle. In most organisms, it is cyclin abundance that oscillates in the cell cycle, not Cdk abundance. All budding yeast cyclins activate the one Cdk (Cdc28), albeit to different extents, and they also help the Cdk to find the right substrates in the cell cycle. Note that the existence of multiple cyclins was already evident from Hunt's original paper when he saw on his gels two oscillating bands in clam eggs. The multitude of cyclins also explains why they were not in Hartwell's *cdc* collection. Cell cycle kinetics may be affected in cells lacking any one of the cyclins, but the cells are not arrested in the cell cycle.

4.3.1.1 G1 Cyclins

At this point, as we realize the variety of cyclins, it is worth taking a little detour and discuss how budding yeast came to be the first organism where G1 cyclins were identified. After the success with the *wee* mutants in fission yeast, an analogous scheme for the identification of small-sized mutants of budding yeast was done. One of these mutants was about half the size of wild-type cells (78, 79). Years later, when the gene was cloned, it encoded a stabilized cyclin, which functioned in late G1. The same gene was also found independently at the same time, looking for dominant mutations that would not arrest at Start by mating pheromone (80). In that case, too, the mutant allele encoded a stabilized version of the cyclin protein, now called Cln3 (Figure 4.11).

Here was a cyclin protein (Cln3), which when stabilized (in *CLN3-1* cells) or overexpressed (in *4xCLN3*+ cells) altered size homeostasis, accelerated Start, and propelled cells to DNA replication. It seemed then that this protein could turn the master switch on, taking the G1 cells out of an MPF-low state, committing them to initiate the process of dividing. Why then were *cln3* loss-of-function mutants not in the *cdc* collection? Well, by now,

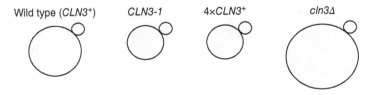

Wild type (*CLN3*+)   *CLN3-1*   *4xCLN3*+   *cln3Δ*

Figure 4.11 The G1 cyclin Cln3 alters the critical size at Start. The dominant *CLN3-1* allele encodes a stabilized Cln3 protein.

you should not be surprised to find that budding yeast has two more of those cyclins (Cln1,2). Only if all three (Cln1,2,3) are deleted cells are arrested at Start, looking like most *cdc28* loss-of-function mutants do (81). More suppressor screens yielded two more cyclins (Clb5,6), with roles a little later in the cycle, in DNA replication, but when present in a high dose, they could resurrect the triple *cln* mutant, raising the total cyclin count in budding yeast to nine. Why stop with yeast genes? Are there any human genes that could provide Cln-like functions and revitalize *cln*-deficient yeast? There are, and that is how some human G1 cyclins were discovered (82). By the mid-1990s, the full complement of cyclins and Cdks was available for most organisms. The master switch comes in many different flavors.

4.3.2 Back to wee1

When we discussed Paul Nurse's *wee1* mutant, we left it as a recessive negative regulator of the Cdk. How does it do so? We now know that there are multiple layers of control of Cdk activity. The Cdk may phosphorylate multiple target proteins to change their function, but it is also phosphorylated by other kinases, conserved in various species. Different kinases target the Cdk/cyclin complex. Wee1 is one of them. Wee1 phosphorylates the Cdk when bound to the cyclin. Wee1 phosphorylates the Cdk on a tyrosine residue in the ATP binding site (Y15 in the human Cdk) and inhibits Cdk activity. Wee1 is a writer of inhibitory modification. Losing Wee1 removes the cells' ability to inhibit Cdk, and they enter mitosis prematurely, at a smaller size. The same effect, small size, is observed either when the targeted tyrosine cannot be phosphorylated, by converting it to an alanine (83), or when the phosphatase (Cdc25) that normally erases the inhibitory Y15 phosphorylation is present in abundance (84). Let's summarize:

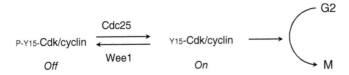

Figure 4.12 Diagram of the G2/M network in fission yeast, based on the phosphorylation status of Cdk at a tyrosine residue (Y15 in fission yeast) targeted by the Wee1 kinase and the Cdc25 phosphatase.

Anything that removes the inhibitory Y15 phosphorylation (e.g., loss of the writer Wee1 or gain of the eraser Cdc25) accelerates mitosis (Figure 4.12).

We have already seen two control levels of Cdk activity (the activating cyclin binding and the inhibitory Y15 phosphorylation), and they will not be the last. Why do cells do that? The more layers of control (with regulators of the regulators and so on), the finer the control level. For a switch of life-or-death decisions, would you want it any other way?

5 Controlling the Master Switch

- Cyclin binding
- Cdk regulation by phosphorylation
- Other proteins that bind Cyclin/Cdk
- Substrates of Cdk activity
- How Cdk targets its substrates

We have pinpointed Cdk activity as the master cell cycle switch. As we saw with the discovery of MPF in the early embryonic cell cycles, changes in mitotic Cdk activity and progression through mitotic landmarks go hand-in-hand. The questions we must answer now are how the activity of the switch is adjusted and how cell cycle events and milestones are linked molecularly with the activity of the master switch. The Cdk switch does not operate as a simple, binary on/off system. Instead, as the cells progress through the cell cycle, Cdk activity gets higher and higher, targeting the substrates relevant for each cell cycle transition. It is like a dimmer switch, except that as it lets more current to go through, a different set of lights is lit, in a precise order.

We will focus on nonembryonic cell cycles, with a G1 phase and cell growth requirements. Based on what we have described so far, a reasonable expectation is that total Cdk activity will be low in cells that do not divide and higher in cells that do. This expectation has been validated

Two from One: A Short Introduction to Cell Division Mechanisms,
First Edition. Michael Polymenis.
© 2023 John Wiley & Sons Ltd. Published 2023 by John Wiley & Sons Ltd.

in various ways, including more recently with Cdk biosensors, enabling Cdk activity monitoring in live cells as they divide (85, 86). In dividing cells, Cdk activity drops when mitotic cyclins are destroyed and cells exit mitosis, as we saw for MPF. Cdk activity stays low in the G1 phase and even lower in quiescent G0 cells until it rises in late G1. Cdk activity continues to increase until cells exit mitosis when it drops again. Before we go into more detail about the relevant targets, we need to look into general ways that cells regulate Cdk activity: binding to cyclins, phosphorylation of Cdk, and association with other proteins.

5.1 Cyclins in Cdk Complexes

During the cell cycle, even in cells with just one Cdk (e.g., in yeasts, also called Cdk1), different cyclins are associated with the Cdk at different cell cycle points. Based on when they peak in abundance and their known roles, the cyclin/Cdk complexes are usually grouped in the following classes:

- G1-Cdk (Cln3/Cdc28 in budding yeast; cyclin D/ Cdk4,6 in humans), peaking in late G1.
- G1/S-Cdk (Cln1,2/Cdc28 in budding yeast; cyclin E/ Cdk2 in humans – sometimes G1/S-Cdk is grouped with G1-Cdk).
- S-Cdk (Clb5,6/Cdc28 in budding yeast; cyclin A/ Cdk2,1 in humans).
- G2/M-Cdk (Clb3,4/Cdc28 in budding yeast)
- M-Cdk(Clb1,2/Cdc28 in budding yeast; cyclin B/ Cdk1 in humans).

While the abundance of Cdks is constant in the cell cycle, that of most cyclins is periodic. Cyclins are unstable proteins, reflected in their low abundance among cellular proteins (87). Typically, targeted protein degradation mechanisms govern cyclins' disappearance in the cell cycle. In contrast, cell cycle-dependent gene expression (usually at the transcriptional level) accounts for the timing of each cyclin's appearance. We will look at these mechanisms in more detail later, but a common theme is that cyclins themselves often set up the temporal pattern in other cyclins' expression. For example, the G1-phase cyclins in yeast trigger the expression of the B-type cyclins, and the S-phase cyclins enable the accumulation of M-phase ones.

Cyclin is necessary for Cdk activity because it changes the Cdk's conformation and allows it to bind its substrates (76). Without cyclin, a part of the Cdk called the T-loop and a small helix next to it block the active site of the Cdk, and peptide substrates cannot access it (Figure 5.1, left). Upon cyclin binding, the T loop moves

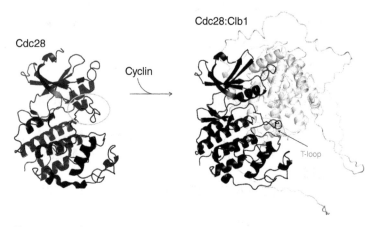

Figure 5.1 The predicted structure of the budding yeast Cdk (Cdc28) monomer (left) was displayed using the coordinates made available from the AlphaFold Monomer v2.0 pipeline (88, 89). The gray circle highlights the T-loop and the small nearby helix that block access to the active site. The predicted structure of the Cdc28:Clb1 (M-Cdk) complex was also displayed using the corresponding coordinates generated in (90), and obtained by permission. Notice the displacement of the T-loop and the adjacent helix in the Cdk:cyclin complex (right).

and the obstructing helix changes to a strand, leaving the active site partially open to substrates (Figure 5.1, right). Cyclin binding also changes the conformation of the ATP binding site of the Cdk, allowing the ATP to bind correctly. Cdk is a proline-directed kinase. Once active, Cdk will phosphorylate serine and threonine residues on target proteins with a full consensus sequence of [S/T*]PX[K/R], or sometimes with a minimal consensus motif ([S/T*]P). All cyclins bound to the Cdk will cause such activating conformational changes. However, the degree of activation is not the same (Figure 5.2). Cyclins impart the "dimmer" properties of the Cdk switch (91, 92). Against a full consensus substrate, the activity of cyclin/Cdk activity rises in the order of expression of the cyclins in the cell cycle (G1/S-Cdk < S-Cdk < G2-Cdk < M-Cdk).

That Cdk sequentially reaches increasing activity thresholds at various cell cycle points has consequences for correctly ordering events in the cell cycle. It will be discussed in detail later, along with how cyclins can also

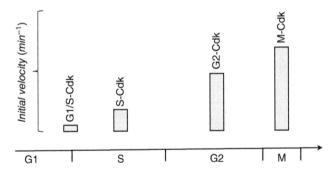

Figure 5.2 Cyclin/Cdk activity toward a peptide containing a full consensus phosphorylation motif rises in the order of expression of cyclins in the cell cycle. The kinetic analysis was performed using a short peptide substrate [PKTPKKAKKL] that only interacts with the Cdk active site and not with cyclins. The values of initial velocity shown were the ones observed when the substrate was added at 400 μM.

Source: Figure adapted with data from Kõivomägi et al. (91).

interact directly with some Cdk substrates, and how cyclins may take the Cdk to specific locations in the cell where the substrate is. Hence, there are multiple ways that cyclins determine the repertoire of substrates that Cdk reacts with. Overall, since a Cdk not complexed with a cyclin has no catalytic activity, the oscillating abundance of cyclins is a crucial mechanism regulating Cdk activity. But it is not the only one, and it is not enough for activating the Cdk.

5.2 Cdk as a Target of Phosphorylations

5.2.1 Activating Phosphorylation

Cyclin binding to Cdk does not fully activate Cdk. All Cdks need to be phosphorylated at a threonine residue (T160 in human Cdk2) in the T-loop for full activation (Figure 5.3, right). The kinases that do this are called Cdk-activating kinases (CAKs). CAK-dependent phosphorylation is conserved from yeast to humans, but CAKs themselves are not. Yeast CAK targets the Cdk monomer, while human CAK activates only cyclin/Cdk dimers. What is perplexing about CAK's activating phosphorylation is that it is constitutive

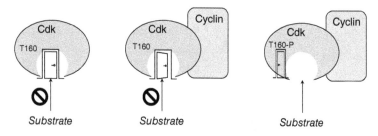

Figure 5.3 Cdk activation by cyclin binding and phosphorylation of Cdk at a T-loop threonine (T160 in human Cdk2) structurally rearrange the active site and "open the door" for full activity against substrates.

throughout the cell cycle. T160 phosphorylation is necessary for full Cdk activity, but this level of control is not used in any way to regulate Cdk activity temporally in the cell cycle.

5.2.2 Inhibitory Phosphorylation

We have already seen the inhibitory phosphorylation of Cdk by Wee1 at Y15 (Figure 4.12). Phosphorylation at Y15 is conserved from yeast to humans. The residue is in the ATP-binding site, explaining the inhibitory outcome. Another inhibitory Cdk phosphorylation in animals is that of the neighboring residue (T14) by a different kinase, whose inhibitory effect is additive to that of Y15 phosphorylation. In all eukaryotes, the inhibitory phosphorylations are removed by Cdc25 phosphatase orthologs. Adding and reversing the inhibitory phosphorylations enable cells to turn very rapidly the Cdk switch on at full throttle, ensuring entry into mitosis. When M-Cdk levels start building up, it phosphorylates both Wee1 and Cdc25. But the phosphorylations have opposing effects on the enzyme targets. Phosphorylation of Wee1 by M/Cdk inactivates Wee1 (i.e., M-Cdk inhibits its inhibitor = M-Cdk activation). At the same time, phosphorylation of Cdc25 by M-Cdk activates Cdc25 (i.e., M-Cdk activates its activator = M-Cdk activation). The result is a rapidly amplifying feedback loop toward one outcome: M-Cdk activation (Figure 5.4). Cells pull all the stops to ensure that M-Cdk activity is as high as possible (the switch is not dimmed at all), so that they enter mitosis and there is no going back.

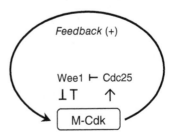

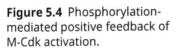

Figure 5.4 Phosphorylation-mediated positive feedback of M-Cdk activation.

5.3 Other Proteins in Cyclin/Cdk Complexes

Once Cdks and cyclins were identified and isolated, it was possible to ask what other proteins may physically associate with them in cells. The logic of this biochemical version of the "guilt-by-association" approach is analogous to the genetic ones we already saw, such as looking for mutants that modify the phenotypes of other mutants. The co-associated proteins fall into two general groups: inhibitors and accessory factors of Cdk activity.

5.3.1 Cdk Inhibitors

Already in the 1980s, antibodies raised against Cdk were used to look for proteins in crude cell extracts that co-precipitated with the Cdk. In budding yeast, a protein (Sic1) was coming down together with Cdk and phosphorylated by Cdk (93). But as cells entered the S phase, Sic1 was gone from the complex. It turned out that Sic1 is a Cdk inhibitor (CKI), but one with a particular taste. It potently inhibits S and M-Cdk (Clb/Cdk) complexes but not G1-Cdk (Cln/Cdk) complexes. As long as Sic1 is around, S and M-Cdk activity would not build up. Unless that is, G1-Cdk activity rises enough in late G1, which then targets Sic1. The G1-Cdk phosphorylation of Sic1 marks it for degradation by ubiquitin-mediated proteolysis. Cells lacking all three G1 cyclins are not viable (Figure 5.5, middle). As we will see, the rise of Cdk activity in late G1 does many things, such as triggering an extensive transcriptional program that includes expression of S phase cyclins. Astonishingly, however, the only essential function of G1 cyclins in budding yeast is to target Sic1 for degradation (94). Targeted proteolysis is of paramount significance in the cell cycle (we already saw how cyclin's destruction is necessary for

CLN1,2,3⁺, SIC1⁺   cln1,2,3Δ, SIC1⁺   cln1,2,3Δ, sic1Δ

Figure 5.5 The only essential function of G1 and G1/S cyclins in yeast is to target the Cdk inhibitor Sic1 for degradation, allowing the activation of S- and M-Cdk.

cells to exit the M phase). Proteolytic mechanisms will be discussed in more detail later. For now, we need to recognize that once the first wave of Cdk activity in the cell rises in G1, its main role is to enable S-Cdk and M-Cdk's next waves by removing their inhibitor, Sic1 (Figure 5.5).

Why do cells need to inhibit S and M-Cdk in G1? After all, as we already mentioned, mitotic cyclins are destroyed when cells exit mitosis, and S and M-phase cyclins are not expressed at high levels in G1. It seems that cells need to be "all-in" at any particular phase of the cell cycle. There can't be even a hint of S or M-phase features in a cell in G1. Their G1 status needs to be stably maintained until G1-Cdk activity rises above a certain threshold, enough to trigger the destruction of Sic1. The role of Sic1 is best described as a "mopping-up" operation, ensuring that no remnants of S and M-Cdk activity are present in a G1 cell. CKIs similar in function (but not in amino acid sequence) to Sic1 exist in fission yeast and flies. They all inhibit S-Cdk and M-Cdk complexes but not G1-Cdk ones, and they all get targeted by G1-Cdk, leading to their destruction at the G1/S transition.

Cdk inhibitors are also used in all organisms as downstream effectors of signaling pathways that regulate the initiation of DNA replication. For example, in addition to Sic1, budding yeast has another CKI, which inhibits G1-Cdk when cells are exposed to mating pheromone. This is how this antiproliferative signaling pathway in yeast causes a Start arrest. As we will see next, analogous mechanisms are widespread in animal cells, with CKIs serving as downstream effectors of various pathways that impact the key commitment step for initiation of DNA replication.

5.3.1.1 Cip/Kip Proteins

In animal cells, the Cip/Kip protein family of CKIs recapitulates some but not all of Sic1's properties. They bind and inhibit G1/S-Cdk and S-Cdk (but not M-Cdk complexes), interacting with both the cyclin and Cdk subunits of those complexes. Unless these CKIs are destroyed, S-Cdk activity will never reach a threshold needed to initiate DNA replication. The "mopping-up" operation of these CKIs is evident in flies that carry loss-of-function mutations in a fly ortholog of one of the Cip/Kip proteins. As fly embryonic cell cycles come to an end, the CKI delays the initiation of DNA replication and helps to impose a G1 phase in the cell cycle. In mutants, the embryo initiates one extra round of DNA replication before it finally terminates its embryonic cell divisions, presumably because enough S-Cdk activity was around in the absence of the CKI (95). Multiple kinase pathways that trigger G1/S in animal cells phosphorylate these CKIs and mark them for degradation. As was the case with Sic1, G1-Cdk complexes are not inhibited by the Cip/Kip inhibitors. But unlike Sic1, Cip/Kip proteins bind G1-Cdk and may even promote G1-Cdk complex formation.

5.3.1.2 INK4 Proteins

These proteins are a separate class of CKIs in mammals. They are specific inhibitors of the G1 Cdk monomer (Cdk4,6). INK4 Cdk inhibitors are major downstream effectors of antimitogens, which generally increase INK4 protein levels. Hence, INK4 proteins reinforce a G1 state in cells. Since the loss of CKIs (INK4 and Cip/Kip) is associated with the initiation of DNA replication, loss-of-function mutations in the genes encoding them are associated with cancer. Several CKIs were also identified genetically as tumor suppressors.

5.3.2 Cks1

Cks1 is a protein that binds cyclin/Cdk complexes, and it is considered a part of the Cdk holoenzyme. It was identified by both genetic and biochemical approaches (96). In high dosage, the *CKS1* gene suppressed loss-of-function Cdk mutants in budding yeast. The Cks1 protein was also found in active cyclin/Cdk complexes. Cks1 proteins are conserved in eukaryotes. They positively affect how Cdk interacts with substrates, as we will describe in more detail later in this chapter.

5.4 What Are Its Targets and How Cdk Phosphorylates Them

5.4.1 Defining the Cdk Substrate Universe

For decades, H1 histone had been used as a substrate to assay Cdk activity, for lack of any alternatives. When I was a postdoc in the late 90s, in seminar after seminar on a cell cycle topic, speakers would invariably say that while there was no question that cyclin/Cdk activity was the master switch of the cell cycle, it was not clear how it was orchestrating cell division, because there were precious few Cdk substrates known. A maestro without an orchestra. A few years later, all that was about to change.

The problem was not unique to Cdk. How do you identify direct substrates, which only fleetingly interact with the kinase, and they are often targeted by more than one kinase in cells? The problem is amplified by the pervasiveness of phosphorylation as a way to modify proteins and control their function. For example, there are >500 kinases in human cells, and one-third of the proteome is phosphorylated. Within the Cdk class, some are not involved in cell cycle control. Furthermore, Cdks are not the only proline-directed kinases. Mitogen-activated protein kinases and stress-activated protein kinases are also proline-directed. Because of such significant overlap among kinases and their specificities, purely genetic approaches to identify substrates are of limited use.

Kevan Shokat devised a creative and transformative solution to the kinase substrate problem. Replacing the endogenous kinase of interest with an engineered version that can accept a bulkier ATP analog did the trick (97). The engineered, analog-sensitive (as) kinase version could use the bulky ATP analog that fits into the enzyme's artificially enlarged binding pocket (Figure 5.6). But all other kinases could not. To find the proverbial needle (direct substrates of the engineered kinase) in the haystack (all other phosphorylation targets in the cells), all you had to do was to use $[\gamma\text{-}^{32}P]$-labeled bulky ATP (Figure 5.6). In crude cell extracts, the only radiolabeled proteins would be those phosphorylated by the kinase of interest. Furthermore, the engineered kinases could also be specifically inhibited by cell-permeable inhibitors that bind only to the mutated kinase active site, while the other kinases are immune. Looking for changes in phosphorylation patterns after short treatments (in minutes) of live cells with such inhibitors could also be used to pinpoint substrates of the inhibited kinase, although they could still be indirect ones. More recently, using bulky ATPγS analogs enable the engineered kinase to specifically thiophosphorylate its substrates in a crude lysate. The thiophosphate label adds a chemical tag that can be derivatized further to purify and identify labeled proteins. Both the substrates' identity and the exact phosphorylation site could be easily identified (98).

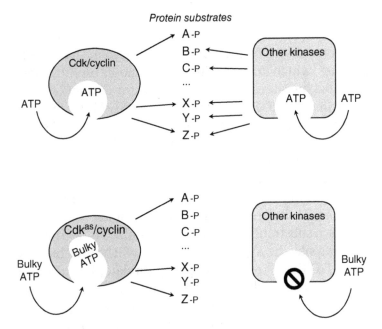

Figure 5.6 Engineered Cdk (Cdk^{as}) that can use bulky ATP analogs and label the Cdk's substrates. Putative substrates would be those shown at the bottom (A, X, Z).

David Morgan's lab pioneered these approaches with Cdk (99–101). Of the ~6,000 proteins of budding yeast, a tenth are phosphorylated by Cdk. At last, we could see the full repertoire of the Cdk effectors, linking Cdk activity with cell cycle-dependent processes. The maestro got her band. We will describe several of the key substrates later, when we focus on specific cell cycle events (e.g., genome duplication and segregation). Until then, let's take a look at how cyclin/Cdk specificity is established.

5.4.2 Cyclin the Recruiter

Once different cyclin/Cdk pairs were found in cells, and at different times in the cell cycle, the most straightforward

prediction was that the cyclins themselves activate Cdk
and help it find the right substrates. With some substrates,
such predictions turned out to be true. Cyclins have docking sites, which interact with specific motifs on substrates.
The docking sites are different among the different classes
of cyclins (i.e., a G1/S cyclin docking site is different from
an S phase cyclin docking site, etc). These cyclin-substrate
interactions occur far from the Cdk active site but help
recruit the substrate to the Cdk. G1, S, G2, and M cyclins
have distinct docking sites, recognizing different substrate
motifs. Even though the intrinsic activity of cyclin/Cdk is
lower in G1 than in subsequent phases of the cell cycle, as
long as a substrate has the motif recognized by a G1 cyclin
docking site, the K_M for that substrate will be lowered, and
the G1-Cdk will efficiently phosphorylate the substrate.
Hence, cyclin works on the kinase by opening the kinase's
active site, and on the substrate, by interacting with it,
increasing its local abundance and the likelihood that the
Cdk will "see" it.

5.4.3 Here Comes Cks1

A related question to the substrate specificity one is the
specificity of Cdk for some sites on a substrate and not
others. Cks1 binds to substrates that have already been
"primed" by initial TP phosphorylation by the Cdk. What
this phosphorylation-dependent docking by Cks1 does is
to promote additional phosphorylations at other sites on
the substrate. As long as other sites are within a certain
distance from the Cks1-mediated docking (and the cyclin-
mediate docking) on the substrate, the Cdk will march
on to phosphorylate them. Imagine now that multisite
phosphorylation is the required outcome for triggering a
cell cycle event. The result of the cyclin and Cks1 docking
interactions with the substrate is that the Cdk may work
processively and complete the phosphorylations without

leaving the substrate. Hence, the Cdk can do its job and achieve the required phosphorylations for biological output even if its intrinsic activity is low (e.g., in G1-Cdk complexes). Due to their docking interactions with some substrates, cyclin and Cks1 could be aptly described as "converter" proteins that produce a graded Cdk activity when the Cdk faces different substrates.

Based on all these experimental observations and the role of docking interactions, Mart Loog recently proposed the model shown in Figure 5.7. A key feature in this model is that Cdk phosphorylations are governed by short motifs encoded in each substrate (Cdk phosphorylation consensus sites, Cks1 docking motifs, cyclin docking motifs, and intervening amino acids that separate these motifs). Because Cdk phosphorylation sites tend to aggregate in the substrates' disordered regions, the coding is essentially

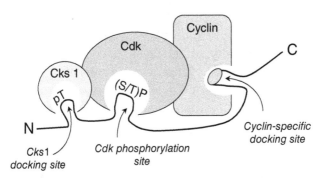

Figure 5.7 The Loog model of how interactions between substrate proteins and components of the Cdk complex determine the rate and specificity of Cdk phosphorylation. A phosphorylated threonine (pT) is needed for interactions between a substrate and Cks1, which then promotes other phosphorylations by Cdk of serine or threonine residues next to a proline. If the substrate has a cyclin docking site it will also be preferentially phosphorylated by the Cdk. All these interactions can be predicted from the amino acid sequence between the amino (N) and carboxy (C) ends of the peptide substrate.

Source: Figure adapted from Örd and Loog (92).

linear and predictable from gazing at the amino acid sequence of the substrate. There is little influence from masking effects due to the three-dimensional structure of the protein substrate. This model explains if and how strongly Cdk will target a substrate, and also when it will be phosphorylated in the cell cycle, as we will see next.

5.5 Ordering Cdk Phosphorylation in the Cell Cycle

To properly do its job, the Cdk master switch must ensure that, among other things, the genome is duplicated before it is segregated (S phase before M) and that both of these processes happen only once in each cell cycle. So far, we have seen that Cdk associates with different cyclins and its activity rises as cells progress in the cell cycle. With help from docking interactions from cyclins and Cks1, some substrates can get phosphorylated even at low or moderate levels of Cdk activity. Added layers of control (CKIs, phosphorylations of Cdk, cyclin degradation) also affect Cdk activity. If any, which one(s) of the above ways contributes to the directionality and order (e.g., S → M) of cell cycle events? Is cell cycle control hopelessly complicated? Or, underneath all these layers, there is a simple core control system, with everything else being add-ons, the proverbial "bells and whistles"? What if you could take out as many of these controls as possible and ask what happens?

5.5.1 Order from Intrinsic Cdk Activity

Paul Nurse's lab did just that in fission yeast. They built a minimal Cdk switch by fusing the Cdk with the mitotic cyclin, expressed as a single polypeptide (102). Furthermore, the Cdk could be an analog-sensitive active site mutant, inhibitable at will by different doses of the specific inhibitors we discussed earlier. The cells carrying this Cdk-cyclin fusion were viable and had the following properties:

- The regulatory and catalytic subunits of the Cdk switch were under the same gene expression control and always present in equal amounts. The regulatory elements that drove its expression were the natural ones that drive expression of the mitotic cyclin. The fusion protein was also sensitive to the same proteolytic mechanisms that destroy cyclin during mitotic exit.
- Separate waves of cyclin abundances did not trigger the rise in Cdk activity in the cell cycle. All the other cyclins with cell cycle roles were deleted, but cells progressed in the cell cycle regardless.
- Binding between the kinase and the cyclin was guaranteed and necessary for Cdk activity. There was no Cdk activity if either the Cdk or the cyclin parts of the fusion were mutated.

How did these cells, expressing only a single Cdk-cyclin complex, manage to initiate and complete S and M phases in the right order? In those experiments, Cdk activity could be adjusted at will by adding different doses of the exogenous chemical inhibitor of Cdk. Natural oscillations in the fusion protein levels due to the elements that normally drive synthesis and degradation of the mitotic cyclin were functionally replaceable with oscillations in Cdk activity enforced by the chemical inhibitor. The only requirement to drive the minimal cell cycle was that the same Cdk/cyclin oscillated between two different activity thresholds: High for M phase, lower for S phase. Lower doses of the Cdk inhibitor blocked DNA replication and higher ones chromosome segregation, suggesting that a lower Cdk activity threshold is required for the S phase than mitosis. Lowering Cdk activity in the M phase, in this case, by adding the Cdk chemical inhibitor, allowed cells to exit mitosis, even on the face of a nondegradable cyclin. If Cdk activity in G2 was artificially made lower and similar to S-Cdk activity, the cells re-initiated DNA replication. Conversely, if G1 cells were suddenly presented with the high mitotic levels of Cdk activity, mitosis was triggered. Regardless of the cell cycle phase cells were in, if Cdk activity were turned full-on, the cells would enter mitosis (with catastrophic consequences). This is analogous to the old fusion experiments, which showed that mitosis is dominant over all other cell cycle phases. Based on this "quantitative" model, it is just the degree of Cdk activation that triggers cell cycle events. As long as the activity keeps rising from S to M, the events will only happen once in the cell cycle and in the correct order.

5.5.2 Precision from Specificity

What about the "qualitative" model then for the specificity endowed by docking interactions between substrates and the Cdk complex? Paul Nurse's lab looked at all Cdk phosphorylations in the cell cycle of cells carrying the single Cdk-cyclin fusion and compared them to the phosphorylation in wild-type cells (103). In both strains, the timing of a substrate's phosphorylation did not correlate with its in vivo dephosphorylation rate. Hence, phosphatase activities in this system do not contribute to the ordering of events in the cell cycle. However, the timing did correlate with the substrate's sensitivity to in vivo Cdk activity. For many substrates, the pattern was the same between the engineered cells carrying only the single Cdk-cyclin fusion and wild-type cells. This explains why the single Cdk-cyclin can support properly ordered cell cycle transitions. However, due to their cyclin-specific docking interactions, other substrates were phosphorylated earlier in wild-type cells than in the single Cdk-cyclin mutant. Hence, although the cyclin-specific docking interactions may not be the main driver for ordering phosphorylations in the fission yeast cell cycle, they fine-tune the proper order.

The single Cdk-cyclin system likely reflects the ancient core properties of the cell cycle switch. The intrinsic Cdk activity is determined by the rate of synthesis and degradation of the regulatory subunit, the cyclin. The Cdk activity rises continuously from the G1/S transition until cells are in the M phase. Every substrate, even poor ones, will be phosphorylated in mitosis when the Cdk intrinsic activity peaks. If they are present, better substrates will be phosphorylated at lower Cdk activity, earlier than the M phase. The high and dominant activity of M-Cdk makes the system exceptionally robust to perturbations and imposes a basic order in cell cycle events. The single Cdk-cyclin system has also been recently tested in budding yeast (104). It turns out that it can support ordered cell cycle progression, albeit a delayed one, and with altered phosphorylation patterns. However, budding yeast cannot survive on it because it cannot properly form a bud. It appears that G1 cyclin specificity has evolved to couple cell cycle progression to an essential morphogenetic event in this organism, bud formation (104). Nonetheless, the fundamental tenet of the "quantitative" model is supported since changes in the intrinsic Cdk activity suffice to order major cell cycle events, such as DNA replication and chromosome segregation.

Why bother then with all the different cyclins, acting at G1/S or S phase? Because cyclins determine both the general intrinsic Cdk activity (low in the early complexes, the highest in M-Cdk), and the apparent activity toward specific substrates that are corralled by cyclin, bringing them to the Cdk. The cyclin-specific docking interactions enable some substrates to be phosphorylated at a lower Cdk intrinsic activity threshold. It also makes the G1/S transition more precise and ensures that late substrates, which still require high Cdk activity, will not be phosphor-ylated prematurely. Interactions that take the Cdk to the right place in the cell also help with the execution of cell cycle events at the right time (105). Hence, it is the cyclin that endows the master Cdk switch with its "dimmer" properties, starting low in S and progressively reaching higher and higher threshold outputs. Which set of lights (i.e., substrates) will the switch light up at each threshold is entirely encoded in the substrates themselves. It depends on both the required intrinsic activity of the Cdk (which is cyclin-mediated) and by any docking interactions that may allow the substrate to bind to the Cdk complex (also mostly cyclin-mediated). It is as if the different intrinsic activity requirements set up a line of substrates waiting to be phosphorylated when their Cdk activity thresholds are reached, but some substrates can jump the line if their sequences allow them to dock with the cyclin.

Adding the Wee1-Cdc25 phosphorylation control of Cdk activity ensures that M-Cdk is activated at the proper time and then reaches its full activity. More add-ons, for example, CKIs, ensure the Cdk is off in G1. But once cells pass through Start in late G1, Cdk activity keeps rising, a necessary feature for ordering the cell cycle. At every higher threshold the Cdk reaches, it fires at its targets in the right order and doesn't go back down until the job is done and cells exit the M phase. Later, we will also discuss how additional surveillance mechanisms or checkpoints sense errors during the cell cycle and delay cell cycle pro-gression until the errors are repaired.

6

A Full Circle of the Switch

- What are the basic properties and requirements for a cell cycle switch?
- Switching from one cell cycle transition to the next
- Proteolysis is part of the switch
- Transcription is another part

6.1 Modeling a Cell Cycle Oscillator

Many types of biological switches and transducers receive some input and produce an output response (106, 107). Before describing the Cdk switch properties that make it so beautiful and ideally suited for the job it has to do, it is useful to review some basic definitions of input (signal, stimulus)–output (response) relationships (Figure 6.1).

- *Linear* response systems. In these systems, plotting the input strength on the x-axis and the output response magnitude on the y-axis will give a straight line. For example, the amount of protein produced by the ribosome (output) is usually linearly related to the amount of mRNA (input), and deviations from this linear response are taken as evidence of specialized translational control (108).

Two from One: A Short Introduction to Cell Division Mechanisms,
First Edition. Michael Polymenis.
© 2023 John Wiley & Sons Ltd. Published 2023 by John Wiley & Sons Ltd.

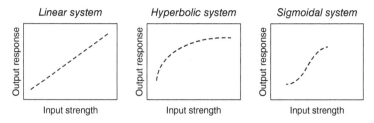

Figure 6.1 Three simple signal-response behaviors.

- *Hyperbolic* response systems. The response is initially linear but then levels off as the system becomes saturated. Examples abound. Think about the Michaelis–Menten relationship that describes an enzyme's activity (output) as a function of the concentration of substrate (input).
- *Sigmoidal* response systems. Here the system responds poorly to small amounts of input. As the input increases, the output response abruptly gets disproportionately stronger and stronger until it reaches a climax. More input does not produce more output (giving an S-type plot when input is plotted against output). Cooperative interactions (e.g., binding of oxygen to hemoglobin) are examples of that behavior. These systems are ultrasensitive because, in the steepest, the middle part of the sigmoidal plot, small differences in the amount of input produce significant output variations.

The sigmoidal, ultrasensitive response is undoubtedly desirable for a cell cycle switch. For example, at low input, you do not want premature activation of cell cycle processes. An ultrasensitive switch can filter out noise. Only if the input presses the button hard enough, the switch will turn on. You also want the cell cycle switch to abruptly generate an output, effectively putting the cell in a new state. This "all-or-none" output is desirable because the cell needs to know if it is dividing or not.

On the other hand, a sigmoidal, ultrasensitive switch is still reversible and sensitive to input changes, just like the linear or hyperbolic ones, depending on the input strength. The button needs to be continuously pressed for the switch to stay on, like a laser pointer's button. Is that desirable for a cell cycle switch? Imagine if the S-Cdk activity is turned on through an ultrasensitive, sigmoidal switch, but then somehow the input falls back down before the cell finishes replicating the genome. The consequences would be catastrophic. Something has to be done to our ultrasensitive cell cycle switch. Instead of being like a laser pointer switch, our cell cycle switch should be like a light switch. Once you turn it on, it stays on even if you do not constantly press the switch. These are more generally called bistable or toggle switches. In our case, we want the cell to have two states, divide (cell cycle switch on) or don't divide (cell cycle switch off). To see how we can get to a bistable irreversible cell cycle switch, we need to look at a few more definitions.

- *Positive feedback.* In this situation, the output is in phase with the input, adding to and making the input larger. For example, a downstream output component (O) affects some upstream input component (I), which produces or activates more output (O) in the future, and on and on in a repeating self-reinforcing loop (I → O → I). Positive feedback loops are common when switching from one state to a new one, in bistable switches.
- *Negative feedback.* In this reciprocal scenario, the downstream output component leads to its inhibition through an upstream input element. As a result, the output counteracts the input's effect, and the system does not switch from one state to another. Negative feedback offers built-in resistance to change, and it is homeostatic. Negative feedback mechanisms are also vital components of

oscillators. The simple case of input (I) leading to output (O), which then inhibits the input (I) (I → O ⊣ I) can generate an oscillator, but not a sustainable one. Instead, the oscillations will be damped with an ever-decreasing amplitude, and the system will reach a steady state.

▪ A built-in *delay* in the output response. *Hysteresis* describes the situation where the value of an output lags behind the changes that caused it, and depends on the prior history of the system's state. It turns out that if such a delay (a new component ⌡) is introduced between the input (I) and output (O) of a negative feedback loop (I → ⌡ → O ⊣ I), then sustained oscillations emerge (107). The time delay in the feedback loop causes a bistable switch repeatedly to overshoot (exceed the target output of the final, steady-state value, i.e., the switch is more ON than it has to be) and undershoot (i.e., the switch is more OFF than it has to be), as it goes back and forth between its two states.

Now we can see what we need to build our cell cycle oscillator switch. The switch should be an ultrasensitive, irreversible, bistable system. We need a sensitive light switch, not a laser pointer-like switch, with a positive-feedback loop to flip it effectively and completely. It should also have a negative-feedback loop with some built-in delay, driving the system back and forth between the two states. Looking at the maturation promoting factor (MPF) switch of the early embryonic cell cycles, abrupt bistability and hysteresis have been experimentally demonstrated (109, 110). Once the Cdk switch is first turned on in late G1, it will stay on until cells exit mitosis, taking several forms (G1/S-Cdk, S-Cdk, M-Cdk, etc.) with different intrinsic activity when it is on. It is time to sketch with broad strokes how each of these forms comes about. We will start with the M-Cdk form, which is probably the ancient, core version of the switch, and work our way around the cell cycle to the other Cdk forms.

6.2 The M-Cdk Switch

6.2.1 Exit from Interphase into Mitosis

Mitotic cyclin transcription peaks in G2 (we will discuss cell cycle-dependent transcription later), and the M-Cdk complex forms at that time, but it is inactive. As we have already seen, the M-Cdk complex is part of several regulatory loops. In late G2, the complex is off because the inhibitory Y15 phosphorylation decorates Cdk1. M-Cdk then phosphorylates and activates its activator, the Cdc25 phosphatase (positive feedback). M-Cdk also phosphorylates and inactivates its inactivator, the kinase Wee1 (double-negative = positive feedback). Such powerful positive feedback generates the two states of the bistable switch (Figure 5.4). The G2 state (M-Cdk and Cdc25 off, Wee1 on) and the M phase (M-Cdk and Cdc25 on, Wee1 off). In some systems, there is even more positive feedback. For example, in human cells, M-Cdk (cyclin B/Cdk1) is in the cytoplasm in G2 but must be in the nucleus to trigger mitosis. The M-Cdk complex phosphorylates cyclin B and promotes the translocation of cyclin B/Cdk1 to the nucleus (Figure 6.2), and on and on. All these positive feedback loops generate an abrupt, all-or-one, robust and irreversible transition from interphase

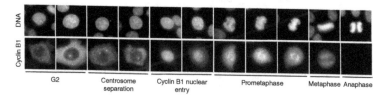

Figure 6.2 Cyclin re-localization in the cell cycle. Immunofluorescence pattern of cyclin B from early G2 phase to anaphase in human HeLa cells. Cyclin B is in the cytoplasm in the G2 phase, enters the nucleus in prophase, localizes on the chromosomes, and is degraded in anaphase (bottom). The staining pattern of the DNA is also shown (top).

Source: [112] Lindqvist et al. (2007), PLoS Biol.

to mitosis (111). But positive feedback alone is not enough for building our oscillator. We still need negative feedback and a time delay mechanism.

6.2.2 The Anaphase Promoting Complex (APC)

Before we see what happens to M-Cdk, we need to describe the role of the anaphase-promoting complex (APC), a large protein complex. The APC is such an integral part of the cell cycle that the whole cell cycle oscillator is often called the Cdk/APC oscillator. When Cdk activity is on (from G1/S to M), APC activity is off, and vice versa. The APC and the Cdk are locked in a perfect yin-yang relation. Each of the two complexes (Cdk and APC) regulates and is regulated by the other. The APC is an E3 ubiquitin ligase. It receives ubiquitin from ubiquitin-conjugating enzymes and attaches it to specific target proteins (Figure 6.3). The giant proteasomes then destroy the ubiquitinated proteins. Unlike other E3 ubiquitin ligases, the APC recognizes different substrates at different times in the cell cycle. The APC

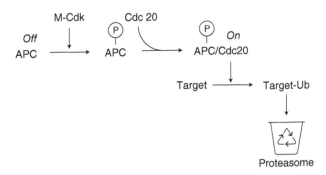

Figure 6.3 M-Cdk activates the APC, which then targets numerous proteins for degradation.

has activators and inhibitors. The APC is inactive in G2 because it cannot correctly bind its substrates. It is the M-Cdk that triggers activation of the APC. Once M-Cdk activity rises, it phosphorylates the APC. The phosphorylated APC can then bind one of its activators, Cdc20, and targets numerous mitotic proteins (Figure 6.3).

6.2.3 From Metaphase to Anaphase

A vital target of the APC/Cdc20 explains its role in "Anaphase Promoting." In metaphase, each of the chromosomes is attached to microtubules emanating from both the mitotic spindle poles. The kinetochore complex of proteins associated with a chromosome's centromere is where the spindle's microtubules attach. The sister chromatids of each replicated chromosome are held together, and they do not separate. The "glue" that holds the sister chromatids together is cohesin complexes. Dissolution of the "glue" and movement of the chromatids to opposite poles defines anaphase. The precise and simultaneous separation of the sister chromatids is arguably the most dramatic morphological landmark of the cell cycle. The lengthways splitting of chromosomes first seen by Flemming in the late nineteenth century (Figure 1.1) immediately suggested that perhaps chromosomes are the basis of heredity. How does the APC/Cdc20 promote anaphase? It destroys securin, an inhibitor of separase, the enzyme that cleaves cohesin. The sequence of events goes like this: APC/Cdc20 ⊣ securin ⊣ separase ⊣ cohesin. The APC waits to do so until all the kinetochores have been properly attached to the mitotic spindle. The "waiting" mechanism is fascinating in itself, but we will discuss it later. It is the result of surveillance mechanisms by the spindle assembly checkpoint, which inhibit the APC until all the kinetochore attachments have been made.

6.2.4 Flipping the M-Cdk Switch Off

Let's see what we have so far regarding our oscillator: A rise of M-Cdk activity, with multiple positive feedback loops that make it abrupt and all-or-none, leading to APC activation, securin degradation, and anaphase. But how does M-Cdk activity go down? If you remember from the discovery of cyclin, cyclin degradation is necessary for a drop in MPF activity. By now, you must have already guessed it. It is the APC that also targets mitotic cyclin for degradation. Destroying the cyclin-activating subunit turns the Cdk switch off, allowing phosphatases to erase the phosphorylations catalyzed by M-Cdk. We got the negative feedback we were looking for: Cyclin B/Cdk1 $\rightarrow$ APC/Cdc20 $\dashv$ cyclin B.

M-Cdk promotes its inactivation. A case of the proverbial "sowing the seeds of its destruction." And not the only one. With the APC-mediated loss of M-Cdk activity, the phosphorylation of APC also goes down, and Cdc20 loses its affinity for the APC. Another co-activator, Cdh1, binds the APC, targeting a host of new proteins, Cdc20 included. APC/Cdh1 continues to be active in G1, destroying any B-type cyclins that come its way. We can now sketch out the major events that set the states of the M-Cdk switch (Figure 6.4).

In most species, the APC/Cdc20-mediated destruction of mitotic cyclin seems to be enough to drive cells out of mitosis. Budding yeast takes no chances and adds more control layers to ensure the desired outcome. First, APC/Cdh1 is also needed during mitotic exit, to fully destroy mitotic cyclins and other proteins. Second, to counter the phosphorylations of the M-Cdk kinase, budding yeast does not just rely on the eventual shift in the equilibrium toward dephosphorylation. It also induces a specific phosphatase to remove them (Figure 6.5). The task falls on the Cdc14 phosphatase, which is sequestered away in

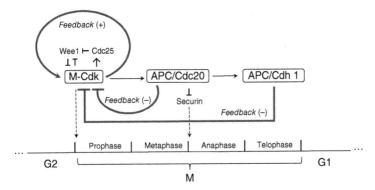

Figure 6.4 Schematic of some major M-Cdk positive and negative feedbacks (shown with thick gray arrows), ensuring that M-Cdk will rise sharply at the G2/M transition and then disappear as cells exit from mitosis.

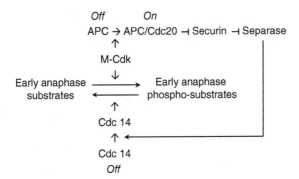

Figure 6.5 The rise of M-Cdk eventually leads to the activation of the Cdc14 phosphatase in budding yeast, which dephosphorylates the early anaphase substrates targeted by M-Cdk. Additional mechanisms activate Cdc14 further, which then dephosphorylates late anaphase substrates, to promote mitotic exit.

the nucleolus until the rise of M-Cdk triggers its release. Another example of M-Cdk setting events in motion that will eventually antagonize it. The activation of Cdc14 is partly triggered by separase (the key event of anaphase), which goes on and dephosphorylates securin, accelerating securin's destruction by the APC. This feedback generates the sharp and synchronous dissolution of cohesin complexes and sister chromatid segregation (113). Cdc14 also accelerates the conversion of APC/Cdc20 to APC/Cdh1. As if all the above were not enough, the lowering of M-Cdk activity as cells exit the M phase allows the build-up of CKIs, the protein inhibitors of Cdk complexes. In budding yeast, Cdk-dependent phosphorylation of Sic1 targets it for degradation, as we have already discussed. With less and less M-Cdk after anaphase, Sic1 levels rise and inhibit the remaining M-Cdk, ensuring that cells exit mitosis (114).

In summary, the rise of M-Cdk triggers positive feedback and pushes cells into the M phase. But the high "Cdk-on" state also starts multiple negative feedbacks that will eventually take cells out of mitosis (APC-mediated destruction of cyclin, accumulation of CKIs, a reversal of mitotic phosphorylations), with the cells attaining a new but stable "Cdk-off" state. Bistability in action. There is also enough built-in delay for our oscillator. Suppose you can measure the cyclin threshold you need to enter mitosis to the threshold you need to exit it. The cyclin threshold "going up" may not necessarily be the same when "coming down." A telltale of hysteresis is if you need to reach a higher cyclin threshold going into mitosis than the cyclin threshold below which you get out. This scenario was experimentally demonstrated for M-Cdk in vitro (Figure 6.6) in frog extracts (109, 110).

Overall, as elaborate as the M-Cdk switch may seem, the design is not unique. The combination of positive and negative feedback is something nature frequently turns to, in other biological oscillators, from the sodium channel-based cellular pacemakers in our hearts to intracellular calcium oscillations and possibly circadian oscillations too (106, 107). The same design also underlies changes in Cdk activity in the rest of the cell cycle.

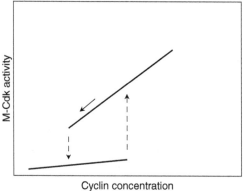

Cyclin concentration

Figure 6.6 Schematic of the ideal behavior of a M-Cdk oscillator. Notice the abrupt changes in M-Cdk activity from relatively small changes in cyclin concentration, and how the cyclin thresholds for turning Cdk on and off are not the same.

Source: Adapted from Pomerening et al. (109).

6.2.5 Unsolved Problem: "Sizing" the M-Cdk Switch

The M-Cdk switch and its various positive and negative feedback loops provide a gratifying view of how Cdk activity changes as cells transition from the G2 phase into and out of the M phase. If we go back to how the Cdk part of the switch was first identified in fission yeast by Nurse (59), we have not answered how *wee* mutants are small. We know that the Wee1 kinase phosphorylates and inhibits the Cdk, but how do cell growth and cell size communicate with Wee1? How do cells know how big they are, and if they have grown enough to signal through Wee1 or other proteins that it is time to get into the M phase? Despite some reports of "sizer" proteins, usually functioning upstream of Wee1 (115), or through Wee1's antagonist, the Cdc25 phosphatase (116), no firm answers to the above questions exist. As we will also see next, links between Cdk1 regulators and mechanisms monitoring cell size remain elusive.

6.3 The G1/S Cdk Switch

At the end of mitosis, the spindle disassembles in telophase, and then it is time for the cytoplasm to divide during cytokinesis. The drop of M-Cdk activity also drives cytokinetic events, which we will discuss later. In the two cells generated after cytokinesis, Cdk activity has been stably turned off. The cells are in the G1 phase, with APC/Cdh1 active (which will destroy mitotic cyclins) and CKI inhibitor proteins. Some cells will stay that way for a long time, even for the rest of their life. If they could only build enough G1-Cdk, could they begin duplicating their genome? Indeed, injecting preformed G1-Cdk protein complexes to quiescent human cells (Figure 6.7), but not S-Cdk ones, is sufficient to trigger DNA replication (117). We will see how G1-Cdk activity eventually rises enough to enable cells to pass through Start or the Restriction point in late G1 and commit to a new round of DNA replication and cell division. We will focus most of our discussion on mechanisms in budding yeast. Animal cells, including

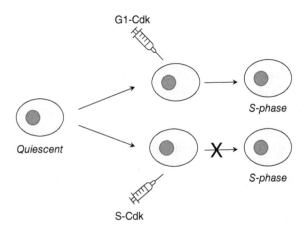

Figure 6.7 Active, preformed G1-Cdk is sufficient to drive cells out of quiescence.

human ones, use similar mechanisms, with the added layer of the INK4 class of CKIs we have already discussed. The INK4 proteins exercise their inhibitory role by binding the Cdk monomer, and they are often the downstream effectors of signaling pathways. A large transcriptional wave characterizes the G1/S switch. We will see how that comes about, and we will also identify the positive and negative feedback loops that we have come to expect from a cell cycle switch.

6.3.1 G1-Cdk Activates G1/S Transcription

6.3.1.1 Doing Away with Transcriptional Inhibitors

The first and critical trigger of the switch is removing an inhibitor of the G1/S transcriptional program. In yeast, that inhibitor is Whi5 (analogous to Rb in humans), inhibiting the SBF transcriptional activator (analogous to certain E2F complexes in humans). In both yeast and human cells, this step is thought to be at the receiving end of growth inputs. Once these inhibitors are removed, cells have completed the commitment step Start or the Restriction point, and they become relatively impervious to growth limitations. How that happens, though, is still mysterious. Figuring out how the switch is triggered in G1/S is critical because it governs cell proliferation rates, as discussed in the second chapter. Overall, some ratio of activators (e.g., G1 cyclins) over inhibitors (e.g., Whi5, Rb) is thought to determine the timing of Start. Conceptually, increasing the ratio of activator/inhibitor could be the result of making more activator or having less inhibitor. In either case, cell growth is proposed to have disproportionate effects on the levels of these regulators: Higher than expected availability of an activator, or lower than expected availability of an inhibitor. We will

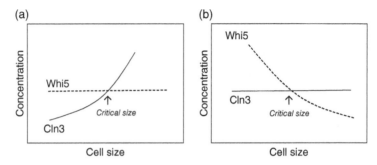

Figure 6.8 Schematic of competing models about how growth inputs control Start in yeast. (a) Cell growth disproportionately produces more of the G1 cyclin Cln3. (b) Cell growth dilutes the Start inhibitor Whi5.

discuss these two models next, and note also that they are not exclusive of each other (Figure 6.8).

6.3.1.1.1 The G1 Cyclin Cln3 Rises

In early G1, Whi5 binds and represses SBF. As Cln3-Cdk activity rises in G1, it targets SBF-bound promoters and activates transcription. Remarkably, Cln3-Cdk does so directly, by phosphorylating RNA polymerase (118). Related, Cdk-like kinases were thought to be the only ones that phosphorylate and activate RNA polymerase. Perhaps Cln3-Cdk reflects an ancient kinase role in both activating transcription and driving cell division, before these roles became functionally distinct. Whi5 then leaves the nucleus, and G1/S transcription is turned on, including mRNAs for the G1/S cyclins Cln1 and Cln2, among many others (119). Except for the direct activation of RNA polymerase by Cln3-Cdk, if you substitute the yeast gene names for the human ones, a similar model emerges for human cells (119). At this point, you should have several questions:

1. How can Cdk activity be around since the inhibitor Sic1 is present, having accumulated from the previous cell cycle's mitotic exit? The answer is that

Sic1 inhibits only B-type cyclin/Cdk complexes. Sic1 does not bind to and cannot inhibit G1 cyclin/ Cdk complexes. This pattern, where an early cyclin is immune to an inhibition imposed on a late-functioning cyclin, is a common theme that we will see again and again (120).

2. Why does Whi5 inhibit the mRNAs' transcription for Cln1 and Cln2, but not the transcription of the mRNA for Cln3? Because *CLN3* transcription is not under the control of SBF. Another reflection of the common theme is that an early cyclin is immune to an inhibition imposed on a later cyclin.

3. Do Cln3 levels even rise in G1? *CLN3* transcript levels do not increase much in G1. An analogous situation is a case for the early-functioning cyclin D in human cells, whose transcription is relatively steady in G1 and not under the control of the E2F transcription factors. Cln3 protein has a very low abundance, and it has been challenging to detect it. Until recently, it was thought that Cln3 protein levels also did not change much in the cell cycle. But, as more sensitive detection methods became available, it is now clear that while newborn daughter cells have little if any Cln3, by mid-G1 and before Start, Cln3 levels rise 7–10 fold (121–124).

4. How do Cln3 levels rise in G1? At present, no molecular explanation exists that altogether accounts for the peak of Cln3 levels before Start. In poorer nutrients, cells have less Cln3 because they repress its synthesis at the translational level (48, 125), and destabilize it even further (126). But all these effects do not necessarily apply to cells going from one cell cycle to the next in the same medium. What is missing is a molecular explanation for the rising levels of Cln3 in G1, under conditions of balanced growth.

6.3.1.1.2 Whi5 Is Diluted Away as Cells Grow in Size

An alternative model centers on the transcriptional inhibitor Whi5 and not on the activating cyclin Cln3. According to this model, Whi5 levels do not follow the overall pattern of biosynthesis in G1. Instead, as cells get bigger, the concentration of Whi5 is gradually lowered until it reaches a threshold below which cells pass through Start (127). In this inhibitor-dilution model, also proposed for Rb, the human Whi5 analog (128), cell growth may not drive synthesis of some activator of cell division (e.g., Cln3 or others). Instead, cell growth dilutes the constant amount of Whi5 the daughter cells received at birth, and this is how cells know how big they are. Notwithstanding subsequent reports that have questioned whether Whi5 is diluted (121, 122, 129), the inhibitor dilution model is an intriguing alternative to activator-based mechanisms.

6.3.2 Positive Feedback at the G1/S Switch

Whichever of the above (or other) models turns out to be correct, what happens next is clear. The G1/S group of transcripts includes numerous mRNAs, many encoding proteins with roles in DNA replication and DNA metabolism. Crucially for our Cdk switch, it also contains the next set of cyclins, the G1/S ones. The little G1/S cyclins produced initially activate more and more G1/S transcription. The transcriptional positive feedback loop generates a switch-like, all-or-none transcription of hundreds of transcripts (130, 131).

However, DNA replication does not begin right away. Cdk inhibitors have to be removed first (Sic1 in yeast, Cip/Kip members in humans). G1-Cdk is immune to these inhibitors, so with the build-up of G1-Cdk activity, Sic1 begins to get phosphorylated, targeting it for degradation.

This time, it is not the APC that does the destroying, but the SCF E3 ubiquitin ligase complex. Loss-of-function mutations in several SCF components lead to late G1 arrest with unreplicated DNA. The critical substrate is Sic1. As we have already mentioned, the destruction of Sic1 is the essential function of G1 cyclins (94). Loss of all G1 cyclins is lethal, but not if Sic1 is also deleted. Among the mRNAs transcribed at G1/S are those encoding S phase cyclins. Although these cyclins are not immune to Sic1, they contribute and accelerate its phosphorylation and destruction. G1-Cdk primes Sic1 for destruction, while S-Cdk finishes it off, generating a sharp, switch-like output at the G1/S transition.

But how can B-type cyclins accumulate, with the APC/Cdh1 lurking in the cell? Shortly after Start, the rise in Cdk activity triggers phosphorylation of Cdh1 and inactivates it because it cannot bind the APC anymore. The S-phase cyclin Clb5 is also immune to APC/Cdh1 and inactivates it, ensuring that later mitotic cyclins can accumulate (132). We see again the theme of an early cyclin's immunity to a block imposed on a later cyclin. With the APC turned off, now the Cdk "dimmer" switch, with new cyclin waves, can reach higher and higher levels of activity. Commitment and moving forward in the cell cycle are ensured.

6.3.3 Negative Feedback at the G1/S Switch

Our switch wouldn't be complete without negative feedback loops. In the true negative feedback form, the S-Cdk generated as a result of G1/S transcription then inhibits G1/S transcription. S-Cdk phosphorylates the G1/S transcription factors (SBF in yeast, certain E2F complexes in humans), which dissociate from DNA, allowing transcriptional repressors to bind to those promoters (119). In yeast,

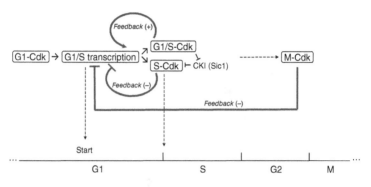

Figure 6.9 The G1/S switch and some of its feedbacks.

M-Cdk also specifically inactivates SBF-mediated gene expression (120). The requirement for Sic1 destruction also imposes a time delay between the transcriptional positive feedback and the negative feedback, for a robust oscillator (Figure 6.9). Lastly, G1 cyclins are phosphorylated and targeted for degradation by SCF. Their job is done; they are not needed anymore. Interestingly, Clb6 (an S-phase, B-type cyclin) is also destroyed early by SCF (the APC destroys all other B-type cyclins in mitosis). As we will see, Clb6 functions in a short window, early in DNA replication.

6.3.4 Physiological Relevance of G1/S Switch in Cancer

In case you were wondering about the significance of the G1/S switch, the commitment step in the cell cycle, soon after their discovery, it became apparent that one or more components of the INK4⊣ cyclin D/Cdk4 ⊣ Rb axis are mutated in most human cancers. I took Table 6.1 from (133), summarizing the percentage of human cancers displaying loss of one INK4 protein (INK4a; by mutation, deletion, or gene silencing), Rb mutation or deletion, and cyclin D or Cdk4 overexpression in different forms of cancer. The table is almost 20 years old, and the numbers are an under-estimate. As of 2020, the latest analysis

Table 6.1 The Rb pathway in human cancer.

Cancer type	INK4a loss	Cyclin D or Cdk4 overexpression	Rb loss
Small cell lung cancer	15%	5% cyclin D	80%
Nonsmall cell lung cancer	58%		20–30%
Pancreatic cancer	80%		
Breast cancer	30%	>50% cyclin D	
Glioblastoma multiforme	60%	40% Cdk4	
T cell acute lymphoblastic leukemia	75%		
Mantle cell lymphoma		90% cyclin D	

Source: Sherr and McCormick (133)/With permission of Elsevier.

of >2,600 cancer genomes by the Pan-Cancer Analysis of Whole Genomes Consortium paints a similar picture (134).

There are now drugs used in the clinic that target this pathway. For example, palbociclib, ribociclib, and abemaciclib are small molecule Cdk4,6 inhibitors (Figure 6.10), receiving FDA approval in 2016–2017 to treat breast cancer (135). These and related compounds are also tested in more than 300 clinical trials (135). From research in yeasts, sea urchins, and frogs, to human patients. What a journey!

6.4 Transcriptional Waves Until the End of the Cell Cycle

We have now examined the states of the G1/S-Cdk switch. The Cdk activity will continue to rise until the G2/M transition, with new cyclin waves. The APC is off, but as we

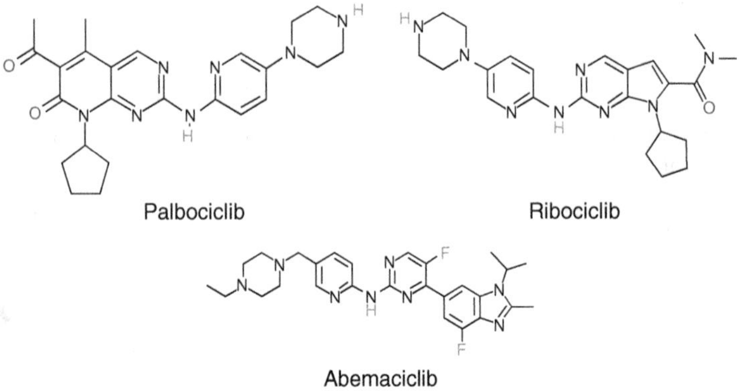

Palbociclib   Ribociclib

Abemaciclib

Figure 6.10 Structures of three FDA-approved inhibitors of Cdk4/6.

Source: The structures were from PubChem (https://pubchem.ncbi.nlm.nih.gov).

have seen, once the M-Cdk state is reached, it will turn the APC on, destroy mitotic cyclins, and cells will exit mitosis. We still need to see how the new cyclin wave comes about.

In yeast, the gradual rise of Cdk activity eventually is enough to activate transcription factors that drive the expression of many gene products with mitotic roles, including the late-functioning mitotic cyclins and the Cdc20 subunit of the APC. The transcriptional wave gets stronger and stronger because increasing M-Cdk drives the expression of more and more cyclin, and on and on. M-Cdk also specifically inactivates G1/S transcription factors (120), so that there is no confusion about the point of the cell cycle the cells are (Figures 6.9 and 6.11). The stage is now set for the M-Cdk switch.

A last transcriptional wave is seen as cells exit the M phase. With the drop of M-Cdk activity, transcription factors that are normally inhibited by mitotic phosphorylations are now activated and drive the expression of gene products that are important for establishing the Cdk-off state, the Cdk inhibitor Sic1 chiefly among them (Figure 6.11).

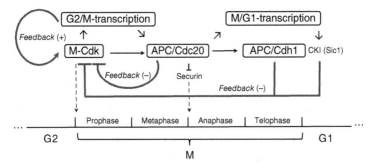

Figure 6.11 Transcriptional networks late in the budding yeast cell cycle.

Overall, the transcriptional network at the G2/M and G1/S transitions in the cell cycle of budding yeast adds more feedbacks and contributes significantly in establishing the different states, as cells enter and exit mitosis (Figure 6.11).

6.5 Comments on Overall Gene Expression in the Cell Cycle

What if you asked a general question about which level of control (transcription, translation, proteolysis) contributes the most to gene expression in the cell? It turns out that transcription accounts for >90%, with translation contributing to about 6–7%, and the rest arising from regulated proteolysis (136). From our discussion of the cell cycle switches, transcription also plays a key role. About 15% of the yeast genes are transcribed in a cell cycle-dependent manner (137). Proteolysis does seem to play an outsized role in cell cycle control compared to other biological processes, at least for the major players. It also probably explains why cell cycle regulators are among the least abundant classes of proteins (87). What about translational control? The answer was difficult to come by because it is much harder to identify translational than transcriptional control. Only in the last few years, with next-generation sequencing methodologies, has it been possible to look at translational control in the cell cycle. In yeast, work from our lab has identified fewer than 100 transcripts that show different translational efficiency in the cell cycle (54–56). We are interested in those targets because they could represent cell growth and protein synthesis inputs, important for cell cycle progression.

Still, it is crucial to recognize that the bulk of gene expression proceeds continuously and with little variation in the cell cycle of yeast (138) and human (139) cells. In yeast, macromolecular synthesis (mRNAs and proteins) shows the expected exponential patterns of growth. Something to contemplate is that within a cell cycle, the ploidy of the cell changes. After the S phase is completed, there are twice as many copies of DNA available for transcription, and then more transcripts for more translation. How the cell manages to adjust its genome content to its size is not known. The problem is related to how cells couple their growth with division we have already dealt with. As we have seen repeatedly (e.g., the lack of information about "sizer" proteins), how growth inputs feed into cell cycle-related processes are some of the least understood mechanisms in the cell cycle.

7 Duplicating the Genome

- How does Cdk control DNA replication initiation?
- Why is DNA replicated once and only once in the cell cycle?
- What happens if DNA is damaged?

Now that we have described the central cell cycle switch and how its activity is regulated during the cell cycle, it is time to link the Cdk activity with key phosphorylation targets and cell cycle events. We will start with DNA replication. Our discussion will not cover the fascinating properties of most DNA replication enzymes in any detail. Instead, we will mostly deal with the critical regulatory steps involving the Cdk complex.

7.1 DNA Replication

In recent years, the DNA replication field has seen the triumphant accomplishment of recapitulating all known control mechanisms in the initiation of eukaryotic DNA replication with purified proteins, in vitro, in a test tube (140). It takes 16 purified complexes in budding yeast, made from 42 distinct polypeptides, to start synthesizing the genome. Thanks to this milestone, the understanding of this landmark cell cycle event is unparalleled.

Two from One: A Short Introduction to Cell Division Mechanisms,
First Edition. Michael Polymenis.
© 2023 John Wiley & Sons Ltd. Published 2023 by John Wiley & Sons Ltd.

To replicate the genome, you have to start from somewhere, an origin. A "landing pad" if you wish, where the enzymes and complexes needed for DNA replication can assemble. The helix will then "melt" and unwind, allowing the DNA polymerases to synthesize the new DNA strands. An *Escherichia coli* bacterium has a single origin in the circular DNA double helix that makes its chromosome. On each eukaryotic chromosome, which contains one long DNA double helix, there are multiple origins. Some "fire" (i.e., initiate new DNA synthesis) early in S phase, and others late. Not all origins necessarily fire in every cell cycle, and in some cases, those that do fire in one cell cycle may not necessarily fire in the next (141). In budding yeast, the origin sequences are short and well defined, but that is not the case for mammalian origins. Because the yeast origins are so well defined, they were used to identify proteins that bind them.

7.1.1 Setting the Stage

Imagine that you want to figure out what activities make the origins fire. A perfectly reasonable experiment would be to take extracts from cells in the S phase, when DNA replication occurs, and ask what fractions of those extracts could make isolated origins fire in a test tube. But all those experiments failed (142). These S-phase extracts were from cells in the Cdk-on state. So why did they fail to support DNA replication? Perhaps the origins needed to be in some different condition, making them competent to fire. Maybe some activity in the extracts prevented them from being capable of firing. It turned out that when Cdk is on in the S phase, it triggers competent origins to fire. But it also prevents the assembly of competent origin complexes on any origins that are not already competent. That can only happen when Cdk is off. So it is only in the G1 phase that origins can get their license to fire. Then, they stay idle until Cdk is turned on, at the G1/S transition, and the S phase begins.

How would you test if a drop of M-Cdk activity is all it takes to assemble a prereplication complex (pre-RC) on origins, making them competent and licensed to fire? Turn Cdk off earlier than it would normally turn off, before anaphase, and see if the pre-RC assembles at origins (143). It did, explaining why in wild type cells origins can only fire once in each S phase. It also showed why before cells can start a new S phase, they must first finish the preceding M phase.

The pre-RC begins assembling on the origins as soon as M-Cdk activity declines, at the end of mitosis. The pre-RC includes the origin recognition complex (ORC), Cdc6, and the minichromosome maintenance (MCM) complex, all of which are ATPases. Once the origins fire, the MCM complex will translocate along the DNA and it will power the DNA strands' separation in all eukaryotes and archaea, a critical function needed for DNA synthesis. To precisely replicate the eukaryotic genome, you must accurately regulate the MCM complex, the replicative helicase motor. The MCM complex is a hexameric ATPase. Two MCM hexamers must be loaded onto DNA, in a stepwise manner (Figure 7.1).

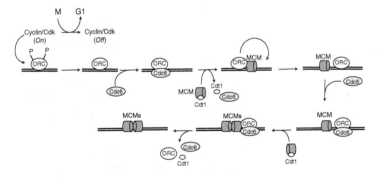

Figure 7.1 Model for MCM loading.

Source: Adapted from Deegan and Diffley (144), Gupta et al. (145).

The ORC complex binds to origins in nucleosome-free regions of the genome and positions neighboring nucleosomes correctly (146). But Cdk phosphorylation of ORC inhibits MCM loading. Once Cdk is turned off at the M/G1 transition, and ORC is dephosphorylated, the ORC recruits Cdc6, and the ORC-Cdc6 platform then recruits the first MCM hexamer. MCM comes along with Cdt1, a loading factor (Figure 7.1). Cdt1 functions as a brace. It holds the MCM bound to ATP and in an open configuration, so that DNA can go through. ORC-Cdc6 will trigger ATP hydrolysis by MCM, closing the MCM ring around DNA, and ejecting Cdc6 and Cdt1 (147). Then, ORC flips and binds DNA on the opposite site of the loaded MCM ring. There it recruits a second Cdc6 molecule, and a second MCM hexamer is loaded, but in the opposite orientation from the first one, with the two MCMs adopting a head-to-head orientation on the DNA (144). If you can picture an MCM complex as a donut with glaze on one side, the way to imagine the origins in G1, is the following: The two MCM donuts are next to each other, interacting through their "glazed" sides, and double-stranded origin DNA going through the donut holes. As we saw, it is MCM itself that "pays" in energy for its loading. But once loaded, the MCMs will not unwind DNA. Instead, the two MCMs sit on the origin and wait, until S-Cdk activity rises (144).

When Cdk rises in late G1, it will phosphorylate Cdc6 and target it for degradation (via the SCF E3 ubiquitin ligase, which also targets the Cdk inhibitor Sic1, as we have seen). This is another way to ensure that the origins will not be licensed to fire again, not before cells exit mitosis and Cdk is turned off (Figure 7.2). Human cells employ an additional mechanism to prevent re-replication until cells exit mitosis. After the origins fire, geminin accumulates. Geminin is a protein inhibitor of Cdt1 and, consequently, of MCM loading at origins. Geminin's levels stay high in the G2 and M phases, but in anaphase, the APC destroys

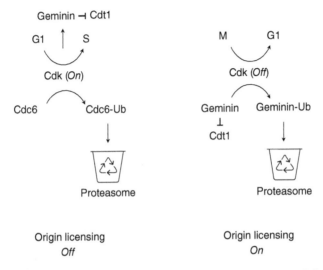

Figure 7.2 Regulation of Cdc6 and Cdt1 (the latter via geminin in human cells) to ensure that origins can only be licensed in the G1 phase of the cell cycle.

geminin (Figure 7.2) and the origins can be licensed again and load MCM (148, 149).

Recently, another example of the connection between replication factors and the Cdk machinery has come to light in budding yeast. As we just discussed, early in the cell cycle, Cdk phosphorylates the licensing factor Cdc6 and targets it for degradation via the SCF ubiquitin ligase system, preventing origin re-licensing. Later in the cell cycle, when S-Cdk drops, any remaining Cdc6 is phosphorylated by M-Cdk, but this phosphorylation does not mark Cdc6 for degradation. Instead, the phosphorylation now primes Cdc6 to be an M-Cdk inhibitor, contributing to the drop in M-Cdk activity and exit from mitosis (150). How can the same process, phosphorylation of Cdc6, by G1/S-Cdk and M-Cdk, lead to different outcomes (destabilization vs. inhibitor of M-Cdk)? Because the interaction of M phase cyclin with Cdc6 is mediated by a specific docking site, and this interaction shields the Cdc6 N-terminal

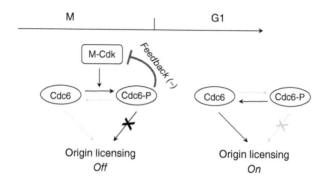

Figure 7.3 The unexpected roles of Cdc6 at the M/G1 transition.

sequences that mark it for degradation. This shielding mechanism is specific to the M-phase cyclin, because the binding of Cdc6 to G1/S-Cdk does not involve the same docking site (150). This is another case of negative feedback at the M-Cdk switch, where M-Cdk triggers its own inhibition. Hence, budding yeast goes over the top and employs not one (through APC activation) but multiple mechanisms (e.g., Sic1 and, now, Cdc6) to ensure that M-Cdk drops and cells exit mitosis. Furthermore, when Cdc6 is bound to M-Cdk, through the specific docking interaction we mentioned, Cdc6 cannot go on to load replication origins (151). That can only happen when M cyclins are destroyed and cells exit M phase and enter G1 (Figure 7.3).

7.1.2 Origin Firing

The S-Cdk complexes trigger origin firing on time. Clb6/Cdc28 activates early-firing origins only, but Clb5/Cdc28 both early- and late-firing ones. The M-Cdk can drive DNA synthesis in the absence of S-Cdk. But even forcing mitotic cyclins to be expressed early, at the same time as S-phase cyclins are normally expressed, the S phase is still delayed. The reason comes down to the specific docking interactions present in S-phase cyclins, which increase their specificity for substrates with DNA replication roles. In human cells, cyclin E/Cdk2 is the major G1/S cyclin, and then cyclin A/Cdk2 is the major S-Cdk.

But how does Cdk make it happen? Suppose you want to identify the critical Cdk substrate. How would you prove it? Well, you could provide the output of the Cdk-mediated phosphorylation, be it a new protein–protein interaction, or the phosphorylation event itself, and ask if that would be enough to trigger DNA replication. Cdks add a phosphoryl group on the hydroxyl group of a serine or threonine on the target protein. Some nonphosphorylated amino acids appear chemically similar to phosphorylated amino acids (for example, aspartate or glutamate for phospho-serine). Introducing such a phosphomimetic substitution in place of a residue on a target protein you suspect to be the critical target of Cdk may bypass the need of the Cdk in the first place. In budding yeast, two proteins passed those tests, Sld2, and Sld3 (152, 153). Their phosphorylation by Cdk promotes critical interactions between other replication proteins. Additional phosphorylations by the DDK complex target MCM. Origin firing converts each double hexamer into two distinct helicase complexes, composed of Cdc45, MCM, and the GINS complex. These are now called Cdc45-MCM-GINS (CMG) complexes (Figure 7.4).

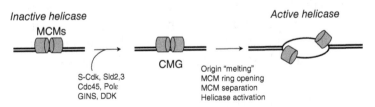

Figure 7.4 Model for MCM helicase activation. Most of the factors that are necessary and sufficient for converting the inactive MCMs to the CMGs are indicated.

Source: Adapted from Deegan and Diffley (144).

Remember that the two MCMs were loaded in opposite, head-to-head orientation at the pre-RC. It was hydrolysis of ATP by the MCMs that enabled their loading, around the double-stranded origin DNA. They stay in a stable ADP-bound state until the origin fires. CMG assembly happens along with some DNA unwinding, and the two CMG complexes are separated from each other, but they remain inactive. More ATP hydrolysis unwinds the DNA further and finally the helicase is activated. After activation, a single strand of DNA goes through each of the donut holes (154). Then each MCM takes off. One of the MCM hexamers is involved in leading strand synthesis and the other MCM hexamer in the lagging strand synthesis, traveling away from each other.

The replisome is one of those awe-inspiring, spectacular cellular machines. It couples the helicase in each CMG complex with the DNA polymerases and other replication factors. The eukaryotic replisome copies 1,000–2,000 bases per minute, with an error rate of about 1 in a billion. DNA polymerase δ is involved in lagging strand synthesis. To some extent, it also participates in leading strand synthesis, but then hands over that job to DNA polymerase ε. After this polymerase switch at the leading strand, DNA polymerase ε, together with PCNA, its processivity factor, reach the upper limits of speed in DNA synthesis (155). Now that we have seen how initiation of DNA replication is coupled to the Cdk system, we need to remind ourselves that chromosomes are not just a long, double-stranded DNA. It is fully assembled sister chromatids that need to be synthesized in S phase and then segregated in mitosis.

7.1.3 Chromatin

All the factors we mentioned above must do their job on chromatin, not naked DNA. It has been known for some time that the chromatin context controls if and when

origins may fire (156, 157). In the in vitro DNA replication system, chromatin prevents nonspecific recruitment
of the MCM helicase. For the replisome to achieve its high
DNA synthesis speed, histone chaperones and nucleosome remodelers disrupt the nucleosome structure. The
nucleosomes are then re-assembled on the newly-made
DNA (158), with a random mix of new and old histone
proteins in them. But the doubling of DNA must be accom-
panied by an increase in histones.

For the most part, the increased demands for histones
are met by a sharp increase in histone protein mRNA levels
in the S phase. The increase in abundance of histone transcripts is complex, and the combined results of increased
synthesis and decreased degradation. In animals, the Cdk
switch has some role in coordinating histone synthesis
with cell cycle progression. G1/S-Cdk phosphorylates and
activates a transcription factor of histone mRNA synthesis
at the G1/S transition (159, 160), but the Cdk control over
histone synthesis is generally less prominent than one
might have expected. Instead, a more basic homeostatic
system is in place. Excess histones are toxic. In yeast, transcription of histone genes is repressed throughout the cell
cycle by a group of repressor proteins, except at the G1/S
transition. If cells accumulate more histones than they
need (e.g., if DNA replication stops or histone gene copy
number is artificially increased), histone transcription is
also repressed.

7.1.4 Sisters Stay Together

As we will describe in the next chapter, the basis of accurate segregation of the genome is that the sister chromatids
of each chromosome generated during DNA replication
stay together until anaphase. Without the "glue" that keeps
the sisters together, the chromosomes cannot be accurately
segregated. We already saw that the M-Cdk triggers the

APC/Cdc20-dependent destruction of securin, the separase inhibitor. In the defining event of anaphase, separase then cleaves cohesin, enabling the sister chromatids to be pulled apart by the spindle. Both the destruction of cohesin in anaphase and its establishment during DNA replication are fundamental and unique aspects of eukaryotic biology.

Cohesin was discovered by Nasmyth and Koshland (161, 162). It is a complex of three proteins that can generate a giant ring structure, wide enough to hold not one but two double-stranded DNA molecules. Two of the proteins are members of the SMC family that have functions associated with genome organization in all organisms. SMC proteins are long and thin, whose association via their hinge domains creates V-shaped heterodimers with ATPase domains at the tips away from the hinge. The two ATPase domains can be joined through another protein, kleisin, closing the complex. It is kleisin that is cut by separase during anaphase.

Based on the above, you probably have the following question: If sister chromatid cohesion in the S phase and its destruction in anaphase are unique to eukaryotes, why are there SMC proteins in prokaryotes? You may have seen the answer in textbooks, showing that the higher-order structure of the DNA in all organisms involves large loops, which are important for its proper packing and organization. Cohesin and the related SMC complexes efficiently extrude through their closed complex the DNA (163). It seems that the core function of these proteins is to pack DNA in all life. Cohesin's role in sister chromatid cohesion looks like an add-on in the course of evolution.

The trimers forming the closed complex we mentioned above cannot do much on their own. Binding by other protein cofactors stimulates ATPase activity and loop extrusion. In nondividing, quiescent cells, cohesin loads onto the chromosomes, making loops. The association of cohesin with chromosomes is dynamic, with the opening and closing of the complex. But once sister chromatid cohesion is established in the S phase (164), it must stay put until anaphase. In the ring model, both sister chromatids go through one cohesin ring (165). In G1 cells, we have one DNA molecule running through each cohesin ring. During DNA replication, sister chromatid cohesion is established in several ways (166). First, the replisome can simply drive through a preexisting cohesin ring on unreplicated DNA (Figure 7.5a). Second, the ring can be transferred behind the replisome in a process that involves replication factors (Figure 7.5b). Third, new cohesin rings can be loaded onto the DNA encircling both sister chromatids (Figure 7.5c).

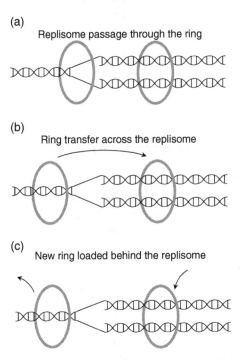

(a) Replisome passage through the ring

(b) Ring transfer across the replisome

(c) New ring loaded behind the replisome

Figure 7.5 Cohesin complexes previously associated with un-replicated DNA converted into cohesive ones behind the replication fork or cohesion built by de novo loading of cohesin molecules onto nascent DNAs.

Source: Adapted from Srinivasan et al. (166).

But as we mentioned, cohesin's association with chromosomes is dynamic, with cohesin opening and closing. On the other hand, cells must stably maintain sister cohesion until anaphase. In other words, the cohesins that trap the two sisters must also stay closed until anaphase. An acetyltransferase associated with the replisome, acetylates cohesin in the S phase, *preventing* the complex from opening (167, 168). An additional protein, sororin, helps maintain sister chromatid cohesion in vertebrates (169).

In summary, sister chromatid cohesion appears to have evolved from universal mechanisms of genome organization. It is timed with replisome progression in the S phase and stably maintained. Until cells reach the metaphase to anaphase transition. Then, the cohesins holding the sisters together will be forcefully and synchronously cut open, enabling the faithful and accurate segregation of chromosomes.

7.2 Checkpoints

7.2.1 The General Concept

We have seen how changes in Cdk activity set in motion events of the cell cycle. Returning to our car analogy to describe a cell "running," if there is a technical malfunction (e.g., blowing a gasket, flat tire, etc.), the car will inevitably come to a stop. Likewise, damage of some kind will probably stop the cell from dividing. But there is a difference in how the car (or the cell) stops. If you drive on the highway and suddenly an engine part explodes, the stop will not be orderly, with likely catastrophic outcomes. If, on the other hand, the malfunction is sensed, an alarm message is transmitted, and the car automatically slows down or comes to a safe stop, then catastrophe is averted.

What if cells have such mechanisms that sense damage or errors and delay or arrest cell cycle progression until the error is corrected? How would you screen for mutants in those mechanisms? Thinking about it in car terms, If no errors or damage occur, having such a faulty sensor system that does not slow you down or stop the car properly won't prohibit you from using the vehicle, and you will notice nothing. Unless damage does occur. Then, the vehicle and you will have no way of knowing, no time to stop safely, and a catastrophic accident will be hard to avoid.

The above considerations were first clearly formulated by Weinert and Hartwell (170, 171). The mechanisms that sense these errors and defects and induce a cell cycle arrest until the defects are repaired were called checkpoints. Weinert and Hartwell proposed that "elimination of checkpoints may result in cell death, infidelity in the distribution of chromosomes or other organelles, or increased susceptibility to environmental perturbations such as DNA damaging agents" (170). As we will see, these predictions are mostly correct. Hartwell and Weinert went a bit further, arguing that checkpoint mechanisms enforce dependency in the cell cycle. At the time, this proposal offered a possible answer to the fundamental question of how cell cycle events are ordered. However, this argument was overly expansive and became harder and harder to reconcile with subsequent results (172). For example, as discussed in detail, the Cdk/APC core control system of the cell cycle is mainly responsible for imposing order and dependency in cell cycle events and transitions. The narrower definition of checkpoints as "surveillance mechanisms that block cell cycle transitions" is more appropriate (172). Over the years, a variety of checkpoints have been proposed to exist. There are currently 169 overlapping "checkpoint" ontology terms (geneontology.org), covering most cell cycle transitions. However, here we will only deal with the DNA damage checkpoint and how it affects DNA replication. In the next chapter, we will also look at the spindle assembly checkpoint.

7.2.2 DNA Damage Checkpoint

The existence of specific mechanisms that delay cell division in response to DNA damage was shown first in bacteria (173). Hartwell and Weinert got to similar mechanisms in budding yeast, reasoning that some mutants sensitive to radiation were so not because they could not repair their DNA but because they could not stop the cell cycle until their DNA was repaired (Figure 7.6). They found the first checkpoint gene, *RAD9*, this way. *RAD9* is not essential, but without it, if there is DNA damage, the cells keep going in the cell cycle, dividing a few times at most and then dying. In contrast, wild-type cells would temporarily arrest in the cell cycle until they fixed the damage, and then continue proliferating (171). The DNA damage response is highly conserved and of immense significance in cancer. For example, the *RAD9* gene of budding yeast and the *BRCA1* human gene (famous for its association

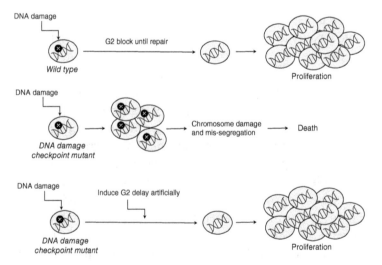

Figure 7.6 Checkpoint mutants fail to arrest the cell cycle upon DNA damage and die after a few cell divisions (middle). Introducing a G2 delay artificially, with mitotic spindle poisons, restores the viability of such mutants, because they are still capable of repairing the damage.

with breast cancer) encode similar adaptor proteins of the DNA damage response.

There are different types of DNA damage (e.g., base modifications, single- or double-strand breaks) from various causes (e.g., spontaneous hydrolysis, radiation, chemicals). Keep in mind that stalled replication forks also trigger the DNA damage response. Here, we will focus on how DNA damage affects cell cycle progression. Before reaching cell cycle components, the signal of DNA damage is transmitted through upstream sensor and effector kinases. The sites of DNA damage recruit and activate the ATR and ATM sensor kinases. Among the targets of these kinases at the damage site are adaptor proteins (e.g., Rad9), which then recruit the next set of effector kinases, Chk1 and Chk2. Phosphorylation of Chk1 and Chk2 at the damage site activates them, and these kinases then leave the damage site and affect the cell cycle machinery. To summarize, a simplified view of the DNA damage response is the following: DNA damage (input) → sensor kinases (ATM, ATR) → effector kinases (Chk1, Chk2) → outputs (including cell cycle block).

The DNA damage response can block cell cycle progression at different points. In human cells, DNA damage will stop progression through the G1/S and G2/M transitions. There are two significant ways that Cdk is inhibited. First, the effector kinases (Chk1,2) drive the accumulation of the inhibitory tyrosine phosphorylation on the Cdk (Figure 7.7). Remember that the inhibitory phosphorylation levels are

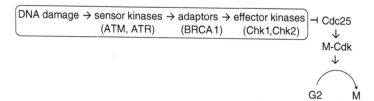

Figure 7.7 Diagram summarizing one example of how the DNA damage checkpoint, shown in the box, arrests the cell cycle at the G2/M transition in human cells.

determined by the antagonism of the "writer" kinase (e.g., Wee1) and the "eraser" phosphatase (e.g., Cdc25). The DNA damage effector kinases (Chk2 in G1, and both Chk1 and Chk2 in G2) phosphorylate Cdc25 isoforms of the eraser, marking them for degradation, enabling the accumulation of the inhibitory phosphorylation on Cdk. The DNA damage response also leads to increased expression of CKIs, especially p21, which also inhibits Cdk complexes (174). However, once M-Cdk reaches high activity, it targets and inactivates effector kinases of the DNA damage checkpoint, ensuring that cells enter mitosis (174).

During the S phase, the replisome itself is an excellent sensor of DNA damage. If it encounters DNA damage, it will stall, triggering the DNA damage response. In budding yeast, the DNA damage response also directly inhibits replication initiation factors. The Chk2 effector kinase (called Rad53 in yeast) phosphorylates and inactivates the DDK kinase and the replication factor Sld3 (175). Hence, no additional origins will fire until the error is fixed. The Cdk-mediated origin firing is blocked by the inactivating phosphorylation of Sld3, which, if you remember, is a crucial Cdk substrate (Figure 7.8). A vital feature of this inhibition mechanism is that Cdk activity stays on. Why would that be beneficial? If you turn Cdk off, the origins that have already fired could be re-licensed, the pre-RC assembled, and re-fire again once the error is corrected. This inhibition of origin firing was one reason that using extracts from cells arrested in the S phase failed to lead to successful reconstitution of DNA replication in vitro (142).

Figure 7.8 Diagram summarizing one example of how the DNA damage checkpoint, shown in the box, prevents further origin firing in S-phase yeast cells.

Prolonged DNA damage response in animals leads to apoptosis or programmed cell death. This explains why chemicals that trigger the DNA damage response (e.g., alkylating agents) are used widely in chemotherapy to initiate such a prolonged response and cell death of cancer cells. Lastly, what about the ends of chromosomes? We will not cover here how telomerase extends the ends of the chromosomes. A proper telomere structure does not trigger the DNA damage response pathways. Unless telomerase is inactivated and the telomeres shorten, leading to profound genome instability and rearrangements.

8 Segregating the Chromosomes

- What is the principle of chromosome segregation?
- What is the mitotic apparatus made of?
- How do the chromatids attach to the spindle?
- How do cells monitor and fix errors?

8.1 Blind Men's Riddle

You may have already heard of the "blind men's riddle" in your undergraduate classes when discussing chromosome segregation. It is a very apt allegory and worth repeating. It goes like this: Two blind men go to a clothing store to buy socks. Each buys his socks, but they like the same colors, so each man buys a gray pair, a blue one, a red one, a green one, a black one, etc. At the register, the clerk puts all the socks in the same bag. How will the blind men manage to get the socks they wanted and paid for? They sit across a table and put the bag in the middle. Then, they take one pair of socks at a time, and each grasps one sock and pulls it. As they pool them apart, one of the men cuts with a knife the thin thread that connects the pair of socks, and each man keeps the sock in his grasp. They go through all the pairs that way, and they end up with one gray pair, a blue one, a red one, a green one, a black one, etc., precisely as they wanted. The socks were accurately

Two from One: A Short Introduction to Cell Division Mechanisms,
First Edition. Michael Polymenis.
© 2023 John Wiley & Sons Ltd. Published 2023 by John Wiley & Sons Ltd.

segregated between them! Substitute the terms of the allegory with the analogous ones from biology, and the principle of chromosome segregation emerges:

- The two socks of a pair → the two sister chromatids of a replicated chromosome.
- The two pairs of socks of the same color → the homologous chromosomes of a diploid organism (e.g., in mammals, one paternal and one maternal).
- The thread that connects the two socks of a pair → cohesin.
- The knife that cuts the thread that connects the two socks of a pair → separase.
- The arms and hands of the two blind men → the microtubules (MTs) of the mitotic spindle.

Keep in mind that in anaphase, all the sister chromatid pairs are separated at once, not one-by-one as in the story, but the requirements are clear: grasp each sister chromatid of a duplicated chromosome from the opposite, different pole (called bi-orientation), pull, and wait for the separase to cut cohesin. That is the crucial point, beautiful in principle and action. It also shows why cohesin is so critical for chromosome segregation. To see how it happens in cells, we need to describe how the spindle is put together and how MTs attach to the chromatids.

8.2 The Mitotic Spindle

As we discussed in the first chapter, when scientists first saw chromosomes in the nineteenth century, they also saw the mitotic spindle. The spindle fibers were sometimes called "thin threads" to distinguish them from the "thick threads," the chromosomes. A beautiful history of the mitotic apparatus studies is given in (176). Mitosis was the first cellular process to be observed under the microscope. Advances in microscopy and advances in understanding mitosis have been locked in a positive feedback loop ever since. The need to better visualize mitosis drove the development of new microscopes, which revealed new mitotic mechanisms, and so on.

At first, however, the spindle fibers were only visible in fixed cells, not live ones. Many argued that those fibers were an artifact of the fixation process. By the 1930s, phase microscopy was developed by Zernike, who was awarded the Nobel Prize in Physics in 1953 for his invention (177). Unlike changes in brightness, phase changes in the light that travels through a cell (or any medium) are not visible to the human eye. By separating the background light from the light scattered from the cell and manipulating them further, phase contrast microscopy reveals foreground details previously unseen in live cells. Additional innovations with polarized light microscopy left no doubt that the spindle fibers and the structures that organize them are live cell processes (176). In addition, these spectroscopic techniques also pointed to the dynamic nature of MTs due to the rapid exchange between spindle MTs and a soluble pool of spindle building blocks.

By the 1960s, fibers 25 nm in diameter were seen in all spindles studied, and the term "MTs" was used to describe these fibers (Figure 8.1a). MTs attach to special structures at the chromosome's primary constriction, the centromere, on each chromatid of a metaphase chromosome. The attachment sites were identified as loci of MT binding and called kinetochores. The spindle is an organized assembly of MTs that attach to chromosomes, trying to pull sister chromatids apart – the arms of the blind men.

8.2.1 Tubulin

What are MTs made of? The most straightforward way to find out is to purify MTs and see their composition,

(a)   (b)

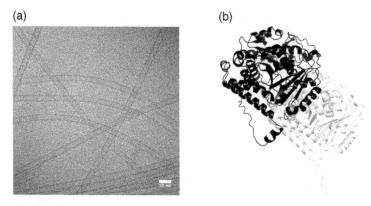

Figure 8.1 The structure of microtubules. (a) Electron cryo-micrograph of unstained frozen-hydrated MTs. (b) Three-dimensional structure of the budding yeast α,β-tubulin heterodimer (Tub1:Tub2). The predicted structure of the complex was displayed using the coordinates generated in (90), and obtained by permission. The α-tubulin (Tub1, colored in white) subunit has a bound GTP, which is not hydrolyzed and is nonexchangeable. The β-tubulin subunit (Tub2, colored in black) also binds GTP, which is hydrolyzed to GDP during polymerization. The plus end of the dimer is at the top.

Source: (a) [181] Manka et al. (2018), Portland Press, CC BY 4.0.

but at the time, it was easier said than done. Later, with the frog extract system, which we saw already in the discovery of MPF (maturation promoting factor), purifying spindles became possible. Another idea is to use a drug that binds specifically at the target of interest (in this case, the spindle), and then use the drug to reel in the drug target. As long as the drug is specific for what you care about, the approach can be successful. A compound that was known to disrupt mitotic spindles was colchicine. Colchicine did not appear to affect other stages of the cell cycle. Because colchicine also bound with high affinity and simple kinetics to cells, Taylor hypothesized that colchicine bound specifically to the building block of MTs and blocked their assembly to the large MT fibers (178). Still, assuming that the drug works as a laser-guided missile, with no "collateral damage" and nonspecific effects, is a risky enterprise and the proverbial "fishing expedition." But all it took was an excited graduate student to isolate the colchicine target, Gary Borysi, whom Taylor warned of the risk in the approach (179).

Borysi and Taylor used radioactively labeled colchicine on various tissues and then looked at what proteins carried the radioactivity. One protein came up and purified to homogeneity, eventually called tubulin. Although the binding was strong in mitotic cells, as one would expect from a spindle component, the most robust binding was from nondividing brain cells. How could that be? We now know that tubulin and MTs are not unique to the spindle. For example, in the "hairs of cells," the cilia, tubulin levels are much higher than any other protein. By the early 1970s, tubulin could be polymerized in vitro to produce MTs (180). The basic unit in the polymer is an α,β tubulin dimer. The two isoforms are very similar (~50% identity, >70% similarity). It turns out that colchicine binds β tubulin, at its interface with α tubulin. At last, there was no question about what the "arms" that pull sister chromatids apart in mitosis were made of (Figure 8.1).

8.2.2 MTs are Dynamic

From both in vitro and in vivo work, it was realized that the two ends of a MT are not the same. In other words, the fibers are polar, having a direction. MTs display dynamic instability, transitioning from fast growth to fast shrinkage. The MT ends distal from the spindle poles were fast-growing, known as the plus-ends (176). To see how this happens, we need to take a closer look at the structure of MTs. To imagine the whole design, think of one strand of tubulin (called a protofilament), with the sequence

$[-(\alpha,\beta)+]n$, with the signs indicating the ends. Then, put 13 such strands side-by-side, in the same orientation, around a hollow cylinder. The overall tubular arrangement is a left-handed helical lattice, with a helical pitch of about 1.5 dimers, with the β-tubulin on the plus end, while α-tubulin is at the minus end (Figure 8.1).

Structure aside, tubulins are not inert bricks. They are enzymes that bind and hydrolyze guanosine triphosphate (GTP). A tubulin dimer in solution is the building block of MTs, and it is in the GTP-bound state. But when the dimer is added to the plus end, the GTP bound to β-tubulin tends to be hydrolyzed. Changing the relative rates of binding to the MT end and GTP hydrolysis underpins the dynamic instability of MTs, continually going through phases of growth and shrinkage. As we already mentioned, the dynamic nature of MTs was suspected long before tubulin was discovered, because of the rapid exchange between spindle MTs and a soluble pool of spindle building blocks (i.e., between free and polymerized tubulin). The polymer vs. free tubulin pools are also similar in abundance (182), reflecting the constant building and re-building of MTs. MTs can go through sudden changes from a growing phase to a shrinking phase (catastrophe) or vice versa (rescue). How many MTs a cell has and the length of these MTs depends on many factors: the concentration of tubulin and the rates of nucleation, polymerization, depolymerization, catastrophe, or rescue. Many proteins that associate with MTs can change one or more of these variables. Some proteins, called motors, use adenosine triphosphate (ATP) hydrolysis to move along MTs. Dynein moves toward the minus end, while kinesin toward the plus end. Other members of the kinesin family move toward the minus end.

The spindle is a beautiful self-organizing polymer (Figure 8.2). The plus ends of some MTs around sister chromatids attach to the chromatids via the kinetochore. In budding yeast, only one MT binds each kinetochore,

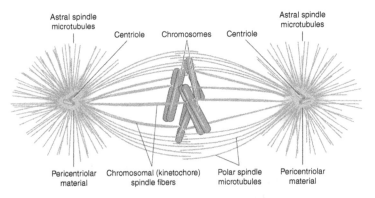

Figure 8.2 Schematic of the mitotic spindle of animal somatic cells. Spindles may also contain short microtubules that do not make contact with either a kinetochore or a spindle pole.

Source: Karp et al. (183)/John Wiley & Sons.

but in animals, there can be 15–35 MTs bound to each kinetochore, crosslinked, and bundled together to a fiber (called k-fiber). These k-fibers are the most stable of all mitotic MTs. An assortment of proteins, including motor proteins, control the dynamic instability of the MTs. They also help orient the minus ends of MTs toward the poles and their plus ends between the poles. The plus ends of some MTs from opposite poles may overlap in the midzone between the poles, and if they are cross-linked, they form polar (a.k.a interpolar) MTs. In animal and yeast cells, MTs from the poles pointing not toward the chromosomes but the cell periphery are called astral MTs, and they help to anchor the spindle to the cell's cortex. Eventually, when each pair of sister kinetochores is attached to MTs from opposite poles, we have the metaphase spindle, arguably the most spectacular structure in the cell (Figure 8.2, and cover photo).

How do we reconcile the instability and dynamic nature of MTs with the seemingly stable scaffold of the mitotic spindle? Mitchison attached a photoactivatable fluorescent probe onto tubulin and fed it to cells. Cells incorporated the potentially fluorescent tubulin into their spindle. Then, he UV-irradiated only a narrow band across the spindle. The tubulin in the MTs within that band got irradiated and fluoresced, but only that tubulin and not the tubulin on either side of the band. Then he watched what happened to the fluorescent tubulin. It traveled to the poles, showing a poleward flux, with MTs moving continuously toward the poles throughout metaphase and anaphase (184). Hence, even if the length of an MT appears constant, the end attached to the pole is shrinking, and the other is growing (Figure 8.3).

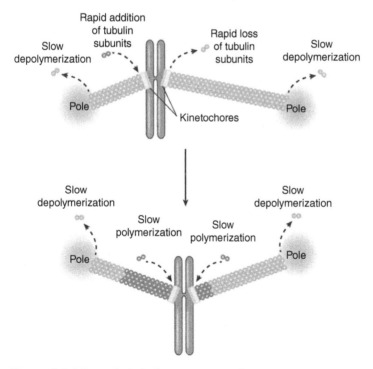

Figure 8.3 Microtubule behavior in metaphase.

Source: Karp et al. (183)/John Wiley & Sons.

As cells progress to metaphase, a chromosome may be attached to MTs from opposite poles that differ in length. Usually, this is adjusted by shrinking the long MT at the kinetochore but growing the short MT at the opposing kinetochore. In metaphase the rates of shrinkage and growth are balanced and the length of the MTs appears constant. These changes are happening in the context of the depolymerization at the poles and the tubulin poleward flux (Figure 8.3).

8.2.3 Scaling the Spindle

Before we describe in more detail the structures at the minus ends of MTs (MT organizing centers, centrosomes) and the chromatid attachment structures (kinetochores), let us ask a few simple questions: Does the spindle need to be of a certain size? How do small cells vs. big cells adjust their spindle size? Do cells "know" how big their spindle needs to be and, if so, how is this done? These questions have attracted a lot of attention, especially in the last decade or so. Several studies have shown that spindle size scales with cell size. The dynamics and position of the spindle also appear to be affected by cell size.

The latest models argue that spindle size in large embryonic cells is closely associated with the total number of MTs in the spindle. More so than the length of individual MTs, their rate of growth, or their lifetime (185). Nucleating new MTs also happens to be a much slower step kinetically than MT growth. How the scaling happens is not clear, but it has been proposed that an inhibitor of MT nucleation is sequestered away at the plasma membrane. The reason for that model is that spindle size, at least in large embryonic cells, scales better with the cell's surface, not its volume (185). Perhaps unsurprisingly, based on our other discussions about size control in the cell cycle, there are many open questions. These include the precise nature of this factor and how these spindle scaling processes are connected with the cell cycle machinery.

8.3 The MT Organizing Centers (MTOCs)

The bipolar spindle is a universal eukaryotic feature, but the pole composition varies. Centrosomes (in animals; Figure 8.4) and spindle pole bodies (SPBs; in fungi) are organizing centers that nucleate MTs at each pole (Figure 8.5). Astral MTs are formed around the centrosomes and SPBs, radiating away from them toward the cell cortex. However,

Figure 8.4 Electron tomogram of centrioles in wing disc fly cells. The cartwheel arrangement of the mother centriole is on top, and the engaged daughter centriole is at the bottom.

Source: [186] Roque et al. (2018), PUBLIC LIBRARY OF SCIENCE (PLOS), CC BY 4.0.

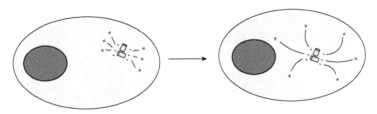

Figure 8.5 Microtubules are nucleated at the centrosome and grow at their plus (+) ends, while the minus (–) ends are at the centrosome. In cells treated with drugs that depolymerize microtubules, the centrosomes will appear without microtubules emanating from them (left), but within minutes after washing the drug away, nucleation of microtubules from the centrosomes will be evident (right).

Source: Adapted from Karp et al. (183).

although plant cell spindles are otherwise similar to those of fungi and animals, plants do not have specialized organelles that nucleate MTs at the poles, nor do they have astral MTs. Animal female meiotic cells also do not have centrosomes, yet they too form beautiful spindles with poles. These observations raise the obvious question of what is the "real" role of MTOCs, to which we will return later.

Possible roles of centrosomes in cell division were noted when chromosomes were first seen under the microscope. Boveri proposed that "the centrosome represents the dynamic center of the cell. . . The centrosome is the true division organ of the cell, it mediates the nuclear and cellular division" (187). Crucially, Boveri also realized that centrosomes were independent and permanent cell organelles that do not disappear when cells complete mitosis. Besides chromosomes, centrosomes and SPBs are the only macromolecular structures that are present in a single copy early in the G1 phase of the cell cycle. Boveri found that during the cell cycle, centrosomes duplicate and separate. They organize the mitotic spindle, determine the cell division plane, and that extra centrosomes cause multipolar spindles and mitotic catastrophe (187).

Each animal centrosome has two tubulin-rich, cylindrical structures called centrioles (Figure 8.4), with additional material surrounding the centrioles. A large assortment of proteins is part of the centrosome and SPBs. The tubulin in centrosomes (and in fungal SPBs) is γ-tubulin, making a ring complex from which MTs can nucleate. As is the case for cells, centrosomes are not generated de novo but only from preexisting centrosomes. In the S phase, two new centrioles are visible, each at right angles next to each of the preexisting centrioles. At the G2/M transition, the two centrosomes separate and move away from each other, driven apart by MT dynamics and associated motor proteins, ultimately forming the two poles of the mitotic spindle. When cells exit mitosis and cell division is complete, each G1 cell has inherited one of the centrosomes.

The duplication of centrosomes and SPBs is thought to be under Cdk activity control, resembling the logic of Cdk control over DNA replication. In yeast, where all the SPB protein components are known, licensing of SPB duplication happens once the Cdk switch is turned off at the M/G1 transition. Activation of Cdk later in G1 triggers SPB duplication and prevents SPB re-duplication until the Cdk is turned off again during mitotic exit. How Cdk limits SPB replication is understood better than how it triggers duplication (Figure 8.6). After mitotic exit, a Cdk target, Sfi1, is dephosphorylated. Sfi1 is on an SPB extension called the "half-bridge." The dephosphorylated Sfi1 promotes the full extension of the half-bridge, which in turn leads to the formation of a satellite structure on which a new SPB will be assembled, next to the previous one when Cdk activity rises again. However, Clb-Cdk also phosphorylates Sfi1, triggering the separation of the SPBs and the formation of the mitotic spindle. But by phosphorylating Sfi1, the SPB cannot duplicate again until the "writer" of the phosphorylation, the Cdk, is turned off, and the "eraser," the Cdc14

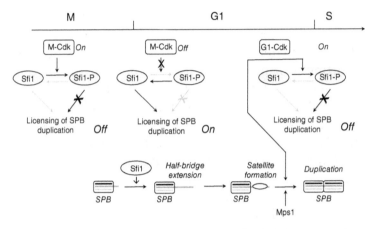

Figure 8.6 Licensing SPB duplication by Cdk is analogous to the licensing of DNA replication. Phosphorylation of Sfi1 by M-Cdk prevents Sfi1 from extending the half-bridge of the SPB and the formation of a satellite structure next to the old SPB. When cells exit M-phase Sfi1 is de-phosphorylated and half-bridge extension and satellite formation are enabled. At the end of the G1 phase, as G1-Cdk activity rises, it triggers SPB duplication (along with another kinase, Mps1). However, Sfi1 is also phosphorylated and the SPB cannot be licensed again until cells exit mitosis and Sfi1 is dephosphorylated.

phosphatase, is turned on (188, 189). Polo kinase, Cdc5 in yeast, also targets Sfi1 and contributes to the above mechanism of licensing SPB duplication.

How Cdk triggers the assembly of another SPB on the satellite structure is not clear. Loss-of-function Cdk mutants arrest with one SPB bearing a satellite. On the other hand, SCF mutants, in which the CKI Sic1 is not degraded and continues to inhibit Clb/Cdk activity, arrest with a duplicated but unseparated SPB (190). Hence, G1-Cdk activity is enough to trigger and complete SPB duplication. Numerous SPB components are Cdk targets. Thus far, there are no examples of mimicking one or more of those phosphorylations by other means and bypassing the Cdk requirement for SPB duplication. But mutating some sites with nonphosphorylatable substitutions impairs SPB assembly and spindle function. For example, mutating the Cdk sites on a core SPB component, Spc42, is lethal (191). Cdk also phosphorylates γ-tubulin, and mutating that site causes mitotic delay and impaired spindle elongation (191). Lastly, another mechanism by which the Cdk switch regulates the SPB cycle is via the expression of key SPB subunits. For example, G1-Cdk triggers the expression of Spc42, as part of the G1/S transcriptional program. Additional kinases, Mps1 in yeast and Plk4 in animals, are also necessary for SPB and centrosome duplication.

Plk4, a polo-like, serine/threonine protein kinase, plays a central role in centriole duplication in animals (192). Plk4 triggers an early structure during centriole duplication (the procentriole) on the surface of the parental centriole, recruiting other centriole biogenesis proteins and γ-tubulin. Artificially increasing the levels of Plk4 triggers the generation of multiple procentrioles at the same time, next to each parental centriole. Conversely, inhibiting Plk4 blocks centriole duplication.

A key question is how centrioles grow to the correct size and length (the cylinder's axis) and how the Cdk oscillator times centriole duplication. In the fly embryonic cell cycles, Plk4 functions as a clock, governing how fast and for how long the daughter centrioles will grow in a cell cycle (193). The activity of Plk4 oscillates and adjusts within a cell cycle to ensure that the centriole grows to its correct size. This is a hallmark of a "sizer" mechanism. The cell "knows" how fast the new centriole is made. If it is growing too fast, it slows it down, and if it has not grown enough, it waits until it does before the next step in the centriole duplication and separation process occurs. These are astonishing new results, arguing that the generation and expansion of self-organizing protein organelles may be governed by similar phenomena we described for whole cells. What about Cdk's role in this process? Here is where another major surprise came up. Although the Plk4 oscillation is normally entrained to the Cdk oscillator, it does not have to. The Plk4 oscillator can run autonomously, on its own, even in the absence of Cdk (194). Perhaps cells use similar homeostatic, "sizer-like" principles to actively monitor and regulate other organelles' biogenesis. Nonetheless, it is still early to tell how general these observations are in other organisms and organelles. But they likely explain why centrioles can assemble in different situations (e.g., cilia), independently of cell cycle progression and the mitotic spindle.

The centriole results suggest that independent oscillators (e.g., the Cdk and the Plk4 oscillator) may be separate but coordinated. Put another way, the Cdk/APC (anaphase-promoting complex) switch may not be the only "game in town" for all cell cycle processes. Our Cdk/APC switch can work together with other oscillators, each put together with similar principles and feedback loops that we discussed. A similar concept had even been proposed for the various transcriptional waves observed in the cell cycle in yeast, coming about "as an emergent property of a transcription factor network that can function as a cell-cycle oscillator independently of, and in tandem with, the Cdk oscillator" (195). However, subsequent work showed that, with a handful of exceptions, the vast majority of temporal, cell cycle-dependent transcription is under "centralized" control of the Cdk/APC switch in yeast (196). As exciting as the Plk4 results in flies are, until more examples and further studies point otherwise, the Cdk/APC switch reigns supreme.

8.4 The Kinetochore

It is time to take a look at how sister chromatids attach themselves to the MTs that will pull them apart from each other. Each eukaryotic chromosome has one and only one functional centromere. Without one (acentric chromosome), the chromosome will not engage properly with the spindle's MTs, and it will not be passed on faithfully to daughter cells. With more than one (e.g., in a dicentric chromosome), a chromosome will likely attach to both poles simultaneously and break. It is often not easy to identify centromeres based on their sequence. Instead, the overall chromatin structure plays a crucial role in establishing a centromere. Specialized histones at centromeres are necessary for the centromere's function to support a kinetochore protein complex formation.

On the specialized centromeric histone H3 variant, called CENP-A in humans, various proteins are assembled to form the inner kinetochore. The specialized centromeric chromatin structure and the inner kinetochore are present throughout the cell cycle. Another group of proteins called the outer kinetochore assembles on the inner kinetochore. The outer kinetochore already assembles in the S phase in budding yeast, but not until the G2/M transition in human cells. It is the outer kinetochore that makes contact with the MTs. To appreciate the kinetochore-MT interactions, imagine that the plus-end of the MT is embedded within the kinetochore. In yeast, a ring-like oligomer of the Dam1 protein will act as a brace around the MT, helped by the Ndc80 complex of the outer kinetochore. Animals do not have the ring-like Dam1 system, but other proteins, the Ska1 complex, mediate attachment to the MTs. Making contact with the MTs emanating from the centrosomes is not the only thing the chromatin-bound kinetochores do. The kinetochores can also nucleate MTs. Even in centrosomes' presence, the kinetochores initiate the formation of some MTs (197). As we will see later, the kinetochore MTs in animal cells have crucial roles in the bipolar attachment of chromosomes.

A radical view of spindle assembly emerged from transformative experiments done by Heald and Karsenti in the

1990s (198, 199). It turns out that even without centro-somes and kinetochores, all you need to provide to frog egg cytoplasmic extracts is chromatin; DNA-coated beads would suffice. MTs assemble around those beads, and with the help of motor proteins, especially dynein, they will form nice-looking bipolar spindles. As long as motor proteins are around, they will sort MT arrangements of random polarity to the standard bipolar spindles. Hence, chromatin, tubulin, and motor proteins seem to be the core requirements for spindle assembly. Other features (e.g., centrosomes) might have been subsequent add-ons, explaining why plant and animal female meiotic cells do fine without them. Centrosomes in animal somatic cells likely tell the cell *where* to form the spindle and not nec-essarily *how* to make one.

8.4.1 Kinetochore-MT Attachment: Stochastic or Deterministic?

Since either the centrosomes or the chromosomes them-selves can nucleate MTs, we are faced with two competing but not mutually exclusive models of how chromosomes attach to MTs. On the one hand, we have the "search and capture" model, formulated in a series of papers by Mitchison and Kirschner (199). MTs from the centro-somes continuously probe their surroundings through their growth and shrinkage cycles until, eventually, a kinetochore manages to capture an MT. This process is mainly stochastic. The centrosomes essentially "flood" the space around chromosomes with dynamic MTs, and it is only a matter of time before one MT bumps into a kinet-ochore. The chromosome's role in this process is rather passive. Alternatively, the short MTs emanating from the kinetochores, but not other MTs nucleated from the cen-tromere in general, interact with spindle or astral MTs, bringing the chromosomes to a "biorientation domain"

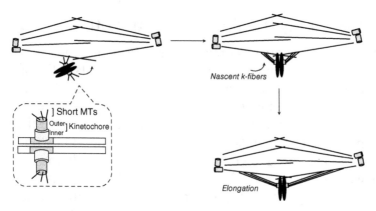

Figure 8.7 The deterministic "mooring" mechanism of chromosome biorientation.

Source: Adapted from Renda et al. (201).

of the spindle, where many crosslinked interpolar MTs are present (Figure 8.7). There, with the help of motor proteins, the short MTs from the kinetochores become nascent k-fibers, which then elongate toward the poles establishing biorientation (200, 201). The latter scenario is a deterministic one, where the chromosome itself grabs onto the spindle instead of the other way around. It is analogous to the mooring process when ships dock at the port. People on the dock typically do not throw ropes in the ship's direction, hoping to hit the target eventually. Instead, it is the crew of the ship that throws mooring ropes (kinetochore MTs) toward the dock (interpolar MTs of the spindle), which are then brought by the people on the dock (motor proteins) to the stable bollards (poles) to secure the ropes and the ship is moored (biorientation). Viewed this way, this "mooring" model is more efficient than the "search and capture" one. It achieves biorientation rapidly. Once one sister kinetochore attaches to the interpolar MTs, the other sister kinetochore will quickly follow (Figure 8.7).

8.4.2 May the Force Be With You

The kinetochore has a massive task: Establish end-on MT attachments, and hold on to those attachments as the MT grows and shrinks, as the end continually changes. For Star Trek fans, the kinetochore must be a favorite macromolecular machine. It senses and responds to the force that pulls the chromatids apart. Why is that important? Both blind men need to grab onto a sock of each pair and pull with some but roughly equal strength for the socks to be divided equally. The goal here is to achieve chromosome congression, where all chromosomes align between the spindle poles, the "equator."

Depolymerization of an MT attached to a kinetochore will result in the chromosome's net pulling toward the pole, as it happens in anaphase, but only if the kinetochore can hold on to a depolymerizing MT. Suppose one sister chromatid binds to an MT that gets shorter due to depolymerization, while the other sister chromatid is attached to an MT that gets longer. In that case, the chromosomes will move up and down the spindle axis during metaphase. Molecular motor proteins contribute heavily to chromosome congression.

Initially, many of the kinetochore associations with MTs are transient and unstable. A network of mitotic kinases (M-Cdk, Aurora-B, Polo, Mps1, and others) and phosphatases (including protein phosphatase 1A and 2A) target kinetochore components, controlling the stabilization of MT attachments. Aurora destabilizes the kinetochore-MT attachment, while Polo stabilizes it. Mitotic kinases also work in converting a lateral to an end-on attachment of the kinetochore to an MT. A key target, but not the only one, of this phosphorylation network is the Ndc80 complex of the outer kinetochore, which is phosphorylated at multiple sites (202). The phosphorylations on Ndc80 and other proteins somehow respond to different levels of MT-generated forces. A correct

kinetochore-MT attachment needs to be stable enough
to withstand the mitotic spindle's strong force (202). Conversely, if the attachment is incorrect, and does not generate tension (e.g., if one or both chromatids are attached
to MTs of the same pole), the kinase network, especially
Aurora-B kinase, will destabilize these MT-kinetochore
attachments. When tension is low, Aurora-B promotes
depolymerization of the MT-kinetochore while maintaining attachment. Aurora-B is localized to the inner
centromere, while its antagonist, protein phosphatase 1,
localizes to the outer kinetochore. Tension is thought to
generate a catch-bond, making the kinetochore-MT interaction more stable. Phosphorylation by Aurora-B probably converts a catch-bond to a slip-bond (203).

A simple model to explain how Aurora-B destabilizes
incorrect MT-kinetochore attachments that do not generate tension is that in the absence of tension, Aurora-
B activity can reach substrates on the outer kinetochore
(e.g., Ndc80) and destabilize the attachment sites (204).
When tension is applied, the distance will increase, and
the Aurora-B activity gradient will not reach the point of
MT attachment as efficiently (Figure 8.8).

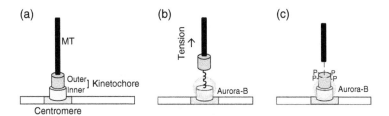

Figure 8.8 Model about how Aurora-B kinase may destabilize
weak MT-kinetochore attachments. (a) Components (MTs,
kinetochore, centromere) involved in chromosome attachment
to the mitotic spindle. (b) Stable attachments generate tension,
pulling the outer kinetochore away from Aurora-B's reach (shown
with a diffuse circle). (c) Unstable attachments do not generate
enough tension to pull outer kinetochore substrates away
from Aurora-B.

There is a particular type of attachment that is incorrect and generates tension. This happens when a single kinetochore is attached to MTs emanating from both spindle poles. These attachments are called merotelic. They account for most chromosome segregation errors and are the primary mechanism of chromosomal instability in cancer cells. It is thought that much effort goes into the "prevention" instead of the "treatment" of merotelic attachments in normal cells. The normal centromere and kinetochore structure are not conducive to merotelic attachments. Mutations in a variety of structural components, including centromeric chromatin, increase merotelic attachments. Treatment with inhibitors of tubulin polymerization (e.g., the popular nocodazole) also massively increases the rate of merotelic attachments. Still, correction mechanisms must exist on top of preventing these attachments because merotelic attachments in anaphase cells are much fewer compared with cells in prometaphase. The k-fibers of the correct attachments are much stronger than the incorrect ones, consistent with the notion that by anaphase, the merotelic attachments are corrected. Aurora-B kinase is again a major player. Its localization to merotelic attachments is very high, and inhibiting its activity increases the rate of merotelic attachments. The closer Aurora-B is to its outer kinetochore substrates, the more likely it is to target them and destabilize the MT attachment. In a merotelic attachment, the MT from the correct pole will pull the attachment site away from Aurora-B, and it will be stabilized. But the MT from the wrong pole comes from behind. It will have to pass over the inner kinetochore, perhaps bringing it within the "strike zone" of Aurora-B (see Figure 8.8). But exactly how the pattern of phosphorylations is translated to precise sensing of tension remains one of the key unresolved questions.

Although the downstream network of mitotic kinases (e.g., Aurora-B) is at the heart of kinetochore-MT attachments, M-Cdk sets in motion mitotic events leading to chromosome segregation. For example, in fission yeast,

M-Cdk phosphorylates a regulatory subunit of the Aurora-B kinase complex, called Survivin. This phosphorylation targets Aurora-B and its regulators (collectively called the chromosomal passenger complex) to the centromere (205). In animals, M-Cdk adds another layer of inhibition of separase, by phosphorylating and inhibiting it. Once APC/Cdc20 is activated, it not only destroys securin, separase's inhibitor, but with the destruction of cyclin, the phosphorylation of separase is reversed, and separase is fully activated, so it can cleave cohesin. The Cdk/APC control system has long tentacles and reaches various aspects of chromosome movement.

8.5 The Spindle Assembly Checkpoint (SAC)

An astonishing aspect of anaphase is the simultaneous cleavage of cohesin. How can it be simultaneous? What stops any bi-oriented pair of sister chromatids from splitting? Enter the SAC. If the APC/Cdc20 is turned on, then cohesin will be cleaved. Hence, the SAC must inhibit APC/Cdc20 until every sister chromatid pair has achieved bi-orientation. The SAC somehow "feels the force" of the MT-kinetochore attachments. If it detects a lack of tension at even a single chromatid, it will keep inhibiting APC/Cdc20. The SAC is a signaling pathway that must sense the absence of tension (input) and, through various steps, lead to an inhibition of APC/Cdc20 so that cells do not exit mitosis (output).

How can you assay the SAC? A classic experiment is to treat cells with chemicals that either reduce (e.g., nocodazole) or increase (e.g., taxol) the MTs' stability, which will arrest cells for hours. Any intervention that shortens the arrest could potentially be involved in the SAC. Much as we saw with the DNA checkpoint mutants, this is how *mad* (mitotic arrest deficient) and *bub* (budding uninhibited by benzimidazole -which depolymerizes MTs) SAC mutants were identified. That an unattached kineto-chore produces a signal that blocks mitotic exit was demon-strated in a beautiful experiment by Rieder and colleagues in the 1990s (206). They zoomed with a laser on the last mono-oriented chromosome that kept the checkpoint on and delayed anaphase. Then, they ablated that chromo-some's centromere, and the checkpoint was relieved.

How the SAC measures tension is unknown. What is better understood is how the SAC blocks the APC (207). As we already saw, Aurora-B phosphorylates various substrates at unattached kinetochores. These phosphor-ylations recruit another kinase, Mps1, the kinase that promotes duplication of the yeast SPB. While the role of Mps1 in centrosome duplication has not been shown in animals, Mps1's involvement in the SAC is conserved in eukaryotes. Once recruited at the unattached kinetochore, Mps1 will phosphorylate additional kinetochore targets, which then recruit SAC components (Bub3, Mad2, Mad3) and Cdc20 to make the mitotic checkpoint complex (MCC) (Figure 8.9). The exact process leading to MCC formation

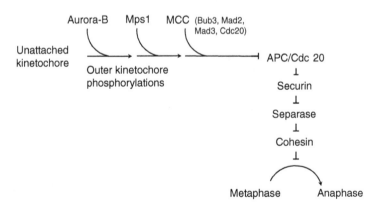

Figure 8.9 Schematic of SAC activation and inhibition of anaphase, through inhibition of APC/Cdc20. MCC is the mitotic checkpoint complex.

is not entirely clear. But in this complex, Cdc20's recognition sites are blocked (208). Once bi-orientation has taken place, the SAC is turned off quickly. New MCCs are not assembled because Aurora-B cannot reach and phosphorylate kinetochore targets that lead to the recruitment and assembly of the MCC. Through incompletely understood mechanisms, the MCCs are also disassembled.

A few closing comments about the SAC. First, merotelic attachments, when the same chromatid is attached to both poles, do not trigger the SAC. Why? Because there is still tension. The lack of SAC involvement might explain why merotelic attachments are the primary cause of aneuploidy. Second, once cells are in anaphase, the SAC cannot be activated. APC/Cdc20 is turned on, and several regulators (e.g., Mps1) are destroyed. Lastly, the medical relevance of the SAC should by now be self-evident. Treatments that trigger the SAC are widely used in chemotherapy. For example, taxol (paclitaxel), which triggers long-term SAC activation because it stabilizes MTs, is used to treat many cancers. As with the DNA replication checkpoint, prolonged arrest triggered by these drugs will lead to programmed cell death.

9

Segregating Organelles and the Cytoplasm

- How do organelles grow in size and number in the cell cycle?
- How do they partition in daughter cells?
- How is organelle biogenesis linked to the Cdk machinery?
- How do cells divide their cytoplasm and split off?

Eukaryotes have a plethora of membrane-bound organelles, each with distinct functions and structures. The amount and kind of intracellular membranes in eukaryotic cells can be staggering. As organelles grow and partition in daughter cells, we need to remember that coordinating membrane (surface) growth with size (volume) growth poses the added complexity of coupling a square function (for surface) with a cube one (for volume). For an ideal, spherical organelle (or, for that matter, whole cell), when volume doubles, the surface area has only increased by about 60%, and the diameter only by 25%. Changing these relations will change the shape of the membrane organelle.

Considering the constraints of the surface-to-volume ratio during cell division has some interesting implications. For example, when one spherical cell divides to give rise to two spherical cells after cytokinesis, a sudden demand of about a quarter to a third of extra plasma membrane surface would be needed to accommodate the same volume.

Two from One: A Short Introduction to Cell Division Mechanisms,
First Edition. Michael Polymenis.
© 2023 John Wiley & Sons Ltd. Published 2023 by John Wiley & Sons Ltd.

These requirements apply for other membrane-bound organelles, the most obvious of which is the large, spherical nucleus, which will become two spherical nuclei in telophase. Where will all this new membrane come from?

To meet the demands for the extra membrane surface, yeast cells upregulate lipogenic enzymes late in the cell cycle (54, 209), and their lipid content increases in mitosis (55). Blocking lipogenesis in yeast and human cells will arrest cells in mitosis (209). In fungi, which undergo "closed" mitosis because the nuclear membrane remains intact, mutants defective in fatty acid synthesis have abnormal nuclear morphology, arguing for the necessary role of lipogenesis in nuclear homeostasis and division (209, 210). Conversely, recent results show that upregulating lipogenesis in yeast accelerates nuclear elongation and division (211), demonstrating that a central metabolic pathway such as lipogenesis can actively drive a cell cycle event instead of just being required for it.

The overall size of most organelles seems to scale with the size of the cell, because a bigger cell will have a greater need for mitochondrial, Golgi, and other organelle functions (212). Much in the same way a large human will have proportionally larger internal organs. These issues are similar to the general problem of coupling growth with division, extended to particular organelles.

The strategies used to segregate organelles seem to differ between animal and budding yeast cells. In animal cells, a common theme is the fragmentation of the organelles to be equally dispersed at random in the daughter cells (Figure 9.1). On the other hand, in budding yeast, organelles are actively transported from the mother cell into the bud, which will eventually become the daughter cell. This reflects the pattern of growth in the budding yeast cell cycle, where after the G1/S transition, most of the cell's growth is directed at the bud. In the next sections, we will look at the morphology, biogenesis, and segregation of some organelles in the cell cycle (213). Lastly, we will close with the division of the cytoplasm and the generation of two cells at cytokinesis.

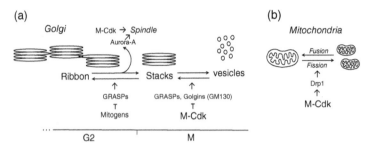

Figure 9.1 In animals, M-Cdk triggers fragmentation of the Golgi apparatus (a) and mitochondria (b), which are then stochastically segregated into the daughter cells.

9.1 The Golgi

The Golgi apparatus is essential for intracellular protein and lipid trafficking and modification. The basic Golgi unit is a sizable flattened vesicle, a cisterna. The higher-order organization of the Golgi is different between mammalian cells and yeast. In yeast, individual cisterna is present throughout the cell. On the other hand, mammalian Golgi is stacks of several cisternae (usually 5–7 per stack). The stacks are further grouped into a ribbon-like arrangement, usually found near the centrosome, in the pericentriolar region. This organization depends on microtubules because microtubule-depolymerizing poisons disperse the Golgi stacks. Until a few years ago, it was thought that mammalian cells could not make a Golgi from scratch. The Golgi could grow only by adding material to a pre-existing Golgi. However, now it seems that individual Golgi cisterna can form de novo (214). This process involves the fusion of vesicles from the endoplasmic reticulum and endosomes into cisternae. Yet, there is more to the story. Some template matrix is still needed (composed of Golgi reassembly and stacking proteins (GRASP) and Golgin proteins) onto which these vesicles organize and adopt their distinctive Golgi identity. The de novo or not organization of the Golgi bears on the Golgi's morphology in the cell cycle.

In mammalian cells, the Golgi fragments at the G2/M transition into thousands of vesicles distributed throughout the cytoplasm. The fragmentation is vital for segregation because each daughter cell will inherit enough Golgi vesicles to reassemble its Golgi stacks. But Golgi fragmentation is essential not only for the Golgi itself but for everything else in the cell. Because blocking Golgi fragmentation will arrest cells in the G2 phase. In case you had any doubts about the significance of organelle growth and division for cell division, you can put those doubts to rest. The question then becomes how Golgi morphology is linked to the cell cycle machinery (215).

The picture that has emerged is a sequential kinase action, leading to the Golgi's step-wise rearrangement: ribbon → stacks → vesicles. The ribbon → stack transition late in G2 is necessary for spindle formation, but the rise in M-Cdk leads to the stack → vesicle transition, necessary for Golgi segregation. First, mitogenic signaling cascades lead to the phosphorylation of GRASP Golgi proteins in the G2 phase. These phosphorylations break the ribbon arrangement so that the stacks are now separated from each other. Ribbon disassembly releases a kinase from the Golgi (a member of the Src family), targeting Aurora-A and phosphorylating it on a specific tyrosine residue. This phosphorylation promotes Aurora-A's recruitment to the centrosome. Aurora-A, together with Cdk, then triggers maturation of the centrosome and spindle formation (215) (Figure 9.1a). This series of events is intriguing because the upstream input is "growth" signaling, mediated by mitogen-activated kinase cascades. This input is then coupled to an organelle's (Golgi) morphology, which acts as an intermediary, for the next steps necessary for a critical cell cycle landmark, spindle formation. In essence, this may be a way to couple cell growth with the cell cycle machinery, with organelles serving as mediators. But the Golgi also benefits because this mechanism is necessary for its segregation into the daughter cells (Figure 9.1a).

As the M-Cdk activity rises, it targets GRASP and Golgin proteins, and the stacks and individual cisterna are broken down into small vesicles (215). These phosphorylations block the usual fusion of vesicles with the Golgi, leading to its fragmentation (Figure 9.1a). The targeting of a Golgi matrix protein, the Golgin GM130, by M-Cdk was one of the first Cdk substrates to be discovered (216). The Golgi vesicles are then more or less equally distributed to the daughter cells just by their sheer numbers. In telophase, the phosphorylations that triggered the disassembly and fragmentation of the Golgi are reversed. The "core" GRASP and Golgin proteins organize and restore the proper morphology, from vesicles → stacks → ribbons, around each of the two nuclei.

9.2 Mitochondria

Mitochondria are a defining and essential organelle of eukaryotic cells. Some have argued more so than the nucleus. After all, the nuclear membrane breaks down in mitosis in mammalian cells (albeit not in fungi), but mitochondrial integrity is never compromised. The essential role of mitochondria is not due to the mitochondrial genome they carry. Yeast and human cells can lose mtDNA. Since mtDNA encodes components necessary for ATP synthesis through the electron transport chain, the mitochondria in these mtDNA-less cells have lost their "powerhouse" attributes. But they can still carry out many other chemical reactions, including the synthesis of iron-sulfur clusters. The latter contribution of mitochondria explains their essential function in all of eukaryotic life (217).

Mitochondria cannot be synthesized de novo. A cell must already have a pre-existing mitochondrion, from which to grow and generate more mitochondria, and eventually, segregate them into daughter cells. At first glance, some

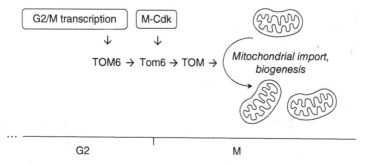

Figure 9.2 In budding yeast, the cell cycle machinery upregulates mitochondrial import and biogenesis late in the cell cycle.

mitochondrial processes may appear analogous to the host's processes and could be under similar cell cycle controls. The obvious example is mtDNA replication. However, mitochondria replicate their DNA with dedicated machinery (using a different mechanism) throughout the cell cycle, not only in S phase. In their long symbiosis with their eukaryotic host, mitochondria "out-sourced" the maintenance of their original genome and synthesis of their proteins to the host, with the vast majority of mitochondrial genes now residing in the nucleus. As a result, the mitochondrial proteins (~1,000 in yeast) are made in the cytoplasm and then imported into mitochondria. It is this import mechanism that now seems to be under cell cycle control in yeast (Figure 9.2).

Mitochondrial proteins have to cross the outer mitochondrial membrane using a protein translocation machine called the TOM complex. In yeast, Cdk activity impinges on the TOM complex in two ways: First, the abundance of the *TOM6* mRNA, encoding a component of the TOM complex, is cell cycle-regulated, rising sharply late in the G2 phase, along with other transcripts (137). Second, Tom6 protein is targeted by G2/M-Cdk (218, 219). The Cdk-phosphorylated Tom6 can then assemble into the TOM complex more efficiently. What's more, Tom6 also promotes more Tom proteins to more efficiently be incorporated into functional TOM complexes. Hence, the more TOM machines the cell has, the more mitochondrial proteins can translocate into mitochondria, and the cell can make more mitochondria.

The significance of the G2/M-Cdk $\rightarrow$ Tom6 $\rightarrow$ TOM $\rightarrow$ mitochondrial biogenesis pathway was tested with the same approaches we mentioned earlier in the coupling of the Cdk switch to DNA replication initiation. Suppose the G2/M-Cdk phosphorylation of Tom6 is necessary for the increased biogenesis of mitochondria in mitosis. In that case, substituting the phosphorylated residue (S16) on Tom6 with a nonphosphorylatable one (a serine $\rightarrow$ alanine change) should block this effect. It did. If the phosphorylation is all you need, a phospho-mimicking glutamate substitution in place of the serine ought to bypass the need for the Cdk phosphorylation. Indeed, the phosphomimetic Tom6(S16E) mutant was imported with greater efficiency and promoted the increased level and activity of the TOM complex. In summary, during yeast mitosis, the Cdk switch opens the "gate" for mitochondrial proteins to make more mitochondria (218). Why in mitosis and not during some other cell cycle phase is not clear.

Mitochondrial shape rarely resembles the sausage-like image seen in many textbooks. Instead, mitochondria are connected (a.k.a. the mitochondrial network) in a reticulum-like structure. Do cells regulate the volume of their mitochondrial network? Remarkably, mitochondrial volume scales with cell volume in yeast, primarily because mitochondria are preferentially segregated in the bud, which will become the daughter cell after cytokinesis (220). The volume of the mitochondrial network in mother cells, on the other hand, keeps getting lower and lower with every cell division of the aging mother cell. How do yeast cells achieve asymmetry in their mitochondrial content between mother and daughter cells? In yeast, mitochondria are transported to the bud from the mother as cargos, using motor proteins. A specialized myosin motor (class V) moves cargos (including various organelles) along the actin cytoskeleton. Interfering with a regulator of this motor eliminated the asymmetric segregation of mitochondria (220).

Although in budding yeast mitochondrial segregation is an active process, until recently it did not appear to be so in mammalian cells. Instead, what we had was similar to the situation with the Golgi (Figure 9.1). In mammals, mitochondria are organized in a network in G1 and S phases, but they fragment in mitosis. Mitochondria fuse during the G1 phase and elongate during S phase, before they fragment in mitosis. Then, the fragmented mitochondria are distributed stochastically in the two cells (221). The mechanism of fragmentation is analogous to that of Golgi fragmentation. Mitochondria undergo constant cycles of fusion and fission. Mitotic Cdk and Aurora-A target and phosphorylate a fission regulator (Drp1), activating it and promoting mitochondrial fission and fragmentation (213) (Figure 9.1b). Upon mitotic exit, Drp1 is targeted for degradation and mitochondria fuse again. Remarkably, recent evidence suggests that human mitochondria during mitosis attach to swirling waves made of actin cables, which help to organize, mix, and segregate mitochondria (222).

9.3 Lysosomes and Vacuoles

The yeast vacuole is analogous to the lysosome of animal cells. In addition to its well-known role in cellular degradation pathways, the vacuole also serves as a storage compartment for low molecular weight compounds. Yeast cells typically have one or few vacuoles. The vacuolar volume is proportional to cell volume. Size scaling is determined by the relative growth rates of the vacuole and the cell (223).

What happens when the vacuolar volume becomes abnormally large? Typically, the size of the cell will also increase. This makes sense because there has to be enough space to accommodate the other organelles and the cytoplasm. A cell with a larger than usual vacuolar volume is "bloated," and this has been used to artificially enlarge yeast cells so that they can be studied using electrophysiology (224). This extreme example raises another issue. In using cell size as a metric of cell "growth," organelle volume changes need to be considered when the normal relations are perturbed. An example is mutants that lack the G1 cyclin Cln3, which are 2–3 times larger than wild-type cells. But these mutants also have more vacuoles and a disproportionately higher vacuolar volume, accounting for some, but not all, of the increase in cell size (225, 226). Cln3 also has a role in maintaining the proper copy number of vacuoles in the cell and their segregation from the mother into the daughter cell (225, 226).

What happens if the daughter cells do not inherit a vacuole from their mother? Vacuole inheritance also relies on motor proteins ferrying them along the actin cytoskeleton to the bud, as we saw with mitochondrial inheritance. But unlike mitochondria that cannot be made de novo, if a daughter cell does not receive a vacuole, it can slowly make one. The interesting part here is that DNA replication will not be initiated until the new vacuole has been made (227). How lacking a vacuole imposes a block on cell cycle progression is unclear. Still, this case provides another example of organelle biogenesis likely serving as a conduit of cell "growth," impinging on the cell cycle.

9.4 Mitotic Fragmentation of the Nuclear Envelope

A hallmark of animal cell division is the breakdown of the nuclear envelope at the G2/M transition. A double membrane bi-layer surrounds the nucleus. The outer membrane is continuous with the endoplasmic reticulum. Underneath the inner membrane, there is a network of mesh-like filaments made from lamin proteins. The lamins and an assortment of associated proteins connect the inner nuclear membrane with chromatin. Nuclear pore complexes (NPCs) are present throughout the nuclear envelope. The rising M-Cdk activity triggers the fragmentation of the nuclear envelope. M-Cdk, together with Polo kinase, phosphorylates NPC components, leading to the disassembly of nuclear pores (213). Likewise, M-Cdk phosphorylates lamins, leading to the breakdown of the mesh-like lamina. Lastly, the spindle microtubules also contribute to the breakdown of the nuclear envelope. The pulling of the chromosomes toward the poles also pulls and tears the nuclear envelope apart. In telophase, with the reversal of mitotic phosphorylations and dissolution of the spindle, the nuclear envelope reforms around each set of chromosomes.

In the "closed" mitosis in fungi, the nuclear envelope does not break down. Unlike animal centrosomes, which are in the cytoplasm, fungal SPBs are embedded in the nuclear envelope. The astral, cytoplasmic MTs radiate toward the cell periphery, while the nuclear MTs toward the chromosomes. Furthermore, there are no lamin proteins. Hence, as the mitotic spindle elongates and stretches, so does the nuclear envelope until it is pinched off in cytokinesis. It is in that last step, with the fission of the elongated nuclear membrane, that a common theme between the "open" and "closed" mitosis emerges. Just within the span of the narrow bridge that connects the segregating nuclei, the NPCs disassemble. The mechanism has been delineated in fission yeast, where a specific protein is responsible for the localized, instead of widespread, NPC disassembly leading to nuclear envelope remodeling and nuclear fission (228). If and how the Cdk switch impinges on this mechanism is not yet clear. Nonetheless, the "open" vs. "closed" mitosis is more a reflection of differences in degree (local vs. global) than in kind.

9.5 Cytokinesis: Two from One

We have already seen plenty of drama in the cell cycle (e.g., think about mitosis). The finale will not disappoint (Figure 9.3). After everything has been duplicated and segregated, it is time to decide where to place, put together, and contract a ring with enough force to cleave the cell (229).

9.5.1 Position

At telophase, there are two nuclei in the cell. It stands to reason that cleavage must happen somewhere between the two nuclei so that each daughter cell inherits one nucleus. But how do cells know where their nuclei are? Different organisms use different strategies. In budding yeast, the cells have already decided much earlier in the cell cycle where to put a bud. The dividing nucleus is pulled through the bud neck in anaphase, and the two nuclei in telophase are on opposite sides of the bud neck. Positioning

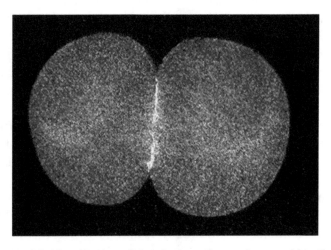

Figure 9.3 Visualization of the cleavage furrow in sea urchin (*L. pictus*) embryos by confocal microscopy, using a fluorescently labeled protein of the cytokinetic contractile ring.

Source: [230] Garno et al., (2021), PUBLIC LIBRARY OF SCIENCE (PLOS), CC0 1.0.

the contractile ring around the bud neck is the obvious choice for this organism. We will not discuss how budding yeast selects where to form a new bud. The mechanisms involved are a model of how eukaryotic cells target their growth to particular areas and directions. Still, polarized growth mechanisms are not used by animal cells to position the cytokinetic ring. Instead, animal cells use the mitotic spindle as a Global Positioning System for *where* to place the contractile ring (Figure 9.4). The role of the spindle in determining the position of cytokinesis is illustrated vividly during animal development, when in some cells the

(a)

(b)

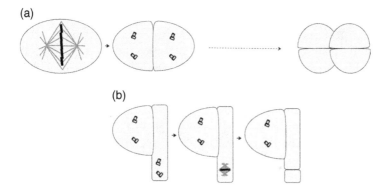

Figure 9.4 Spindle position determines cleavage site.
(a) Typically, once an egg and its descendant cells form the spindle, cytokinesis will follow, with the cleavage furrow forming between the poles. The centrosomes in each cell (shown as cylinders) indicate the spindle poles that have formed before the second division of each cell. (b) In a classic experiment (231), the egg was allowed to divide once to form a two-cell embryo. Then, one of the cells was drawn into a micropipette, causing it to assume an elongated shape, with its centrosomes placed close to each other at one end of the cylinder. The cylindrical cell completed cleavage shortly afterward, whereas its sister cell had not yet begun to cleave. These results suggest that cleavage happens between the spindle poles, regardless of their position. Cleavage occurs earlier in the cylindrical cell because the distance from the poles to the cleavage site is reduced, taking less time for the cleavage signal to reach the surface.

Source: Adapted from Karp et al. (183).

spindle moves and rotates to the desired location, leading to programmed asymmetric cell divisions. Spindle formation also provides an obvious timing cue for *when* to position and assemble the cytokinetic ring (Figure 9.4).

If the spindle is used for positional information, it must somehow communicate that information to the cell cortex, which, after all, is where the ring will be placed. To account for the spindle's role in relaying positional information, two primary models have gained support over the years. In model 1, astral microtubules pointing toward where the equator (or the midzone) of the spindle meets the cell's cortex carry some ring positioning signal. In model 2, interpolar microtubules, which do not attach to kinetochores, generate a cleavage signal in the spindle midzone that reaches the cortex closest to it. The two models are not mutually exclusive, and the evidence argues for a combination of both. Much of the original pieces of evidence came from "mechanical" experiments (229). For example, in cells where the spindle poles had been moved physically close to each other and away from the chromosomes, cleavage occurs between the poles. This result argues that the signal travels through the astral microtubules from the poles to the cleavage site (model 1). On the other hand, if you impose a physical block between the midzone and the cortex, you can ask if and where cleavage occurs. Putting in place such blocks by applying pressure with a blunted needle yields intriguing results (232). Imagine the spindle as a globe, with the two poles at "north" and "south." If you place the block on the "east" of the equator, cleavage will be initiated at the "west," but not at the "east," and vice versa, arguing that a signal from the midzone has to reach the cortex (model 2). Applying the block shortly after (>1 minute) anaphase onset did not affect cleavage initiation, suggesting that the signal had already reached the cortex by that time.

There are several steps in the pathway that lead to contractile ring assembly. We will pick up the pathway from a key activator and go backward to the "trailhead" to see how that activator is linked with the spindle and the mitotic events we have already discussed. The critical activator of the contractile ring assembly is the RhoA-GTPase. How critical? Critical enough that if you artificially activate RhoA anywhere on the membrane and at any point in the cell cycle, it will trigger cleavage (233). RhoA belongs to a GTPase family of proteins that regulates actin dynamics in the cell. When they hydrolyze GTP, all such GTPases have difficulty ejecting the GDP from their active site to bind and hydrolyze another GTP. Hence, they engage with and are activated by guanine nucleotide-exchange factors (GEFs), which replace the GTPase-bound GDP with GTP. The key GEF that activates the Rho-GTPase, in this case, is Ect2. It is primarily through Ect2 that spindle-related processes and factors control the positioning of the contractile ring. What are these spindle-related processes and factors?

The centralspindlin complex, which includes a kinesin motor protein moving toward the plus ends of microtubules, performs vital functions in contractile ring assembly by interacting with lipids at the plasma membrane, and by binding and activating Ect2, the GEF that activates the RhoA-GTPase (229). Now we have a direct physical linkage, through the centralspindlin complex, between the spindle and RhoA-GTPase at the plasma membrane: Spindle MTs (centralspindlin) → Ect2 (GEF) → RhoA-GTPase - - - → cytokinesis. Hence, kinesin motors on interpolar MTs concentrate centralspindlin complex and active RhoA-GTPase where the plus ends of the MTs are adjacent to the cell's cortex (Figure 9.5). But how is this localized activation of RhoA-GTPase linked to the cell cycle machinery?

M-Cdk phosphorylates both centralspindlin and Ect2, inhibiting them. This is a simple way to inhibit cytokinesis until M-Cdk activity drops in anaphase. Indeed, premature inactivation of M-Cdk can trigger cytokinesis (234). In anaphase, additional mitotic kinases (e.g., the Polo kinase)

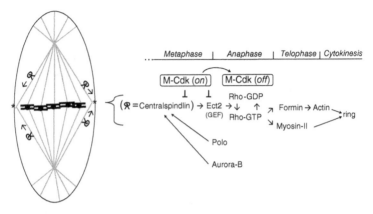

Figure 9.5 Diagram of how the centralspindlin complex from MTs activates Rho-GTPase at the cortex positions ring assembly. M-Cdk inhibits centralspindlin, but Polo and Aurora-B kinases activate it.

phosphorylate and activate centralspindlin, while Aurora-B kinase inhibits an inhibitor of centralspindlin (229). The net outcome of all the above is straightforward: The spindle is required for cytokinesis, but as long as M-Cdk activity is on, cytokinesis is off.

9.5.2 Assemble

As soon as cells exit mitosis, the locally activated RhoA is now ready to orchestrate ring assembly. Let's look at what the ring is made of to understand better how RhoA assembles it. The ring is made of actin filaments and myosin-II (a nonmuscle myosin). Actin filaments are helical, made of two strands of actin. They are also polar, with a barbed (b) and pointed (p) end, based on their microscopic appearance. The b-end grows faster than the p-end, and it is analogous to the plus end of MTs. Unlike MTs, however, actin filaments are thinner, more flexible, and can be arranged in various ways in the cell. Myosin-II also has a filament structure but with motor properties because it has a motor domain at one end. The tail domains of several myosin-II filaments interact to form thick filaments, with the motor domains on the surface on both ends. But the ends are again polar, with the motor domains pointing in the opposite direction. The myosin-II filament does not grow or shrink, but it can generate force. Imagine your torso and your two arms extended from it, pointing in the opposite direction, and capable of pulling on to whatever they grab onto. That arrangement (torso + arms) is the thick myosin-II filament.

The actin filaments serve as the scaffold, or tracks, onto which myosin-II will move and exercise force, pulling the actin filaments together (Figure 9.6). The actin filaments would have to be radially assembled around the cleavage site, with their b-ends anchored at the cortex and their p-ends pointing toward the center. Then, each myosin-II thick filament will bridge two diametrically opposed actin filaments. Each of the two motor ends of the myosin-II filament will "rope-in" the p-end of the actin filament it is attached to, bringing the anchor points (at the b-ends) closer together, pulling the membrane inward. Remarkably, the above picture can be reconstituted in a test tube, in vitro. Helping the "crowding" of actin artificially (with methylcellulose) will lead to rings of actin filaments inside artificial liposomes (235). Adding myosin-II to the mix will constrict the actin rings (229).

The abundance of ring components, actin, and myosin-II is not cell cycle regulated. Although some preformed actin filaments from elsewhere in the cell may be incorporated in the contractile ring, many are nucleated and formed on-site. Active RhoA triggers an actin nucleation factor, formin, leading to actin filament formation, facilitated by yet another protein, profilin, at the cleavage site. The Rho-GTPase also activates a phosphorylation cascade (mediated by two downstream kinases), leading to myosin-II phosphorylation and activation. Hence, RhoA activation tells the cell where to assemble the ring and ensures that the ring's components will form (229).

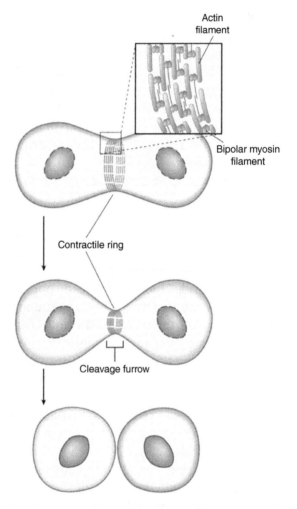

Figure 9.6 The contractile ring during cytokinesis. Actin filaments become assembled in a ring at the cell equator. Contraction of the ring, which requires the action of myosin, causes the formation of a furrow that splits the cell into two.

Source: Karp et al. (183)/John Wiley & Sons.

In budding yeast, another ring of proteins, the septins, line up the inside of the bud neck at the G1/S transition. This step is activated by G1/S-Cdk activity. Then, the same processes and components as in animal cells, Rho-GTPase and the actin-myosin-II complexes, lead to the assembly of the contractile ring.

9.5.3 Contract

The mechanism of tension generation by myosin-II on actin filaments is analogous to muscle contraction. The generated forces have been measured in several cases, reaching ~10–50 nN in sea urchin cells (236). To generate force, the b-ends of the actin filaments need to be anchored properly. The anchoring mechanisms have been studied extensively in fission yeast. Even the number of actin and myosin-II subunits needed to make a ring ~125 nm in diameter and with a circumference of 11 µm have been calculated (229). How exactly the ring generates tension is not entirely understood. It appears that sidewise anchoring of actin b-ends, parallel to the plasma membrane, is the simplest way to generate tension (229). Various proteins likely participate in the anchoring mechanism. Intriguingly, how fast the contractile ring constricts seems to scale with cell size. The bigger the cell, the faster it constricts. As a result, the time it takes to close the ring is similar in small and large cells (229). Another connection between cell size and the cytokinetic ring is also seen in budding yeast. The septin ring's diameter organized at the bud neck at the G1/S transition shows a sizer behavior (237).

A key difference between muscle contraction and cytokinesis is that the contractile ring disassembles as it constricts, although the thickness of the ring stays constant. Most of the available evidence supports the idea that the ring rebuilds itself much faster than it tightens, with the turnover of the ring components measured in seconds (229). Nonetheless, the precise mechanisms that couple constriction and disassembly of the cytokinetic ring are unknown (238).

Lastly, closing and disassembling the ring are not enough. There is vesicle transport to provide the necessary membrane material at the cleavage site. In mammalian cells, toward the end of cytokinesis, just before abscission, when the two cells are cut-off from each other, a transient structure forms between them, called the midbody. The midbody's lipid composition is unique, arguing for specific mechanisms that generate and localize these lipids toward the cytokinetic ring (239). Myosin-II is not present in all eukaryotes, but only in organisms on the branch that includes amoebas, fungi, and animals. In other branches (e.g., plants) that lack myosin-II, cytokinesis is less efficient and does not involve a contractile ring. Instead, cytokinesis relies on membrane traffic and fusion to build a new plasma membrane and cell wall to separate the daughter cells, the same processes that lead to abscission in yeast and animal cells. Hence, the targeted membrane deposition and fusion during abscission seems to be the ancient mechanism of cytokinesis, with the contractile ring assembly and contraction added later in eukaryotic evolution.

We will close by pointing out that the midbody was first described and named by Walther Flemming. The same Flemming that saw and described mitosis in the late nineteenth century. In his honor, the central portion of the midbody is called the Flemming body. Even with the short and incomplete descriptions of the processes we covered, the progress since Flemming's time is dizzying. But it is also stimulating and invigorating for pushing further. Lastly, although the processes we described appear complex at first glance, with numerous factors involved (and a myriad more we did not discuss), there are some simple lessons underneath that complexity. To set things in motion and in the proper order, all you need is the activity of one enzyme (Cdk) adjusted to different levels in the cell cycle. Furthermore, at the most fundamental level, even the most complex mitotic structures (e.g., the spindle or the cytokinetic ring) arise from self-organizing polymers. Perhaps the cell cycle is not hopelessly complex after all. A list of all the protein names mentioned in all chapters is shown in Table 9.1.

Table 9.1 List of proteins or complexes and their function.

Protein or complex	Function	System
CDK machinery		
Cdc28	Cdk (cyclin-dependent kinase)	Budding yeast
Cdc2	Cdk	Fission yeast
Cln3/Cdc28	G1-Cdk	Budding yeast
cyclin D/Cdk4,6	G1-Cdk	Mammals
Cln1,2/Cdc28	G1/S-Cdk	Budding yeast
cyclin E/Cdk2	G1/S-Cdk	Mammals
Clb5,6/Cdc28	S-Cdk	Budding yeast
cyclin A/Cdk2,1	S-Cdk	Mammals
Clb3,4/Cdc28	G2/M-Cdk	Budding yeast
Clb1,2/Cdc28	M-Cdk	Budding yeast
cyclin B/Cdk1	M-Cdk	Mammals
MPF	Complex between Cdk and mitotic cyclin	Animal embryonic cell cycles
Cks1	Processivity factor of cyclin/Cdk complexes	Yeasts and animals
Phosphorylation of CDK		
CAK	Kinase targeting and activating Cdk	Yeasts and animals
Wee1	Kinase targeting and inhibiting Cdk	Fission yeast and animals
Cdc25	Phosphatase targeting and activating Cdk	Fission yeast and animals

(*Continued*)

Table 9.1 (Continued)

Protein or complex	Function	System
CDK inhibitors		
Sic1	Inhibitor of S-Cdk and M-Cdk complexes	Budding yeast
Cip/Kip proteins	Inhibitors of G1/S-Cdk and S-Cdk, but not M-Cdk, complexes	Mammals
INK4 proteins	Inhibitors of G1-Cdk	Mammals
Proteolysis		
APC	Anaphase-promoting complex, an E3 ubiquitin ligase, catalyzes the transfer of ubiquitin to a protein substrate.	Yeasts and animals
Cdc20	Activator of APC, turned on by M-Cdk1	Yeasts and animals
SCF	An E3 ubiquitin ligase, catalyzes the transfer of ubiquitin to a protein substrate	Yeasts and animals
Cdh1	Activator of APC, turned on when M-Cdk1 activity goes down at M/G1	Yeasts and animals
Sister chromatid cohesion		
Securin	Inhibitor of separase, the enzyme that cleaves cohesin	Yeasts and animals
Separase	An enzyme that cleaves cohesin	Yeasts and animals
SMC proteins	Chromosome compaction and organization, family members part of cohesin in eukaryotes	Eukaryotes and prokaryotes

Table 9.1 (Continued)

Protein or complex	Function	System
Cohesin	Complex that tethers sister chromatids together	Budding yeast and animals
Kleisin	Cohesin protein	Budding yeast and animals
Sororin	Helps maintain sister chromatid cohesion	Vertebrates
Cell cycle transcription		
Whi5	G1/S transcriptional repressor	Budding yeast
SBF	G1/S transcriptional activator	Budding yeast
E2F complexes	G1/S transcription	Mammals
Rb	G1/S transcriptional repressor	Mammals
DNA replication		
pre-RC complex (MCM, Cdc6, ORC proteins)	Licensing origins of DNA replication	Yeasts and animals
ORC	Origin recognition complex	Yeasts and animals
Cdc6	pre-RC component	Budding yeast and animals
MCM	Hexameric helicase	Yeasts and animals
Cdt1	Loading factor for pre-RC assembly	Budding yeast and animals
Geminin	Inhibitor of Cdt1	Animals

(Continued)

Table 9.1 (Continued)

Protein or complex	Function	System
Sld2	DNA replication factor	Budding yeast
Sld3	DNA replication factor	Budding yeast
DDK	Kinase that targets and activates DNA replication factors	Budding yeast and animals
CMG complex	Active helicase in DNA replication	Budding yeast and animals
Cdc45	CMG component	Budding yeast and animals
GINS	CMG component	Budding yeast and animals
DNA polymerase δ	Involved in lagging strand synthesis	Budding yeast and animals
DNA polymerase ε	Involved in leading strand synthesis	Budding yeast and animals
PCNA	Processivity factor of DNA polymerase ε	Budding yeast and animals
DNA damage checkpoint		
ATR	Sensor kinase of the DNA damage response	Mammals
ATM	Sensor kinase of the DNA damage response	Mammals

Table 9.1 (Continued)

Protein or complex	Function	System
Rad9	Adaptor protein, linking the sensor and effector kinases of the DNA damage response	Budding yeast
Chk1	Effector kinase of the DNA damage response	Mammals
Chk2	Effector kinase of the DNA damage response	Mammals
BRCA1	Adaptor protein, linking the sensor and effector kinases of the DNA damage response	Mammals
Rad53	Effector kinase of the DNA damage response, analogous to human Chk2	Budding yeast

Microtubules and associated proteins

Tubulin (α & β)	Building blocks of microtubules	Yeasts and animals
Tubulin (γ)	Tubulin that nucleates MTs (e.g., at centrosomes or SPBs)	Yeasts and animals
Dynein	Motor protein, moves toward the (−) end of MTs	Yeasts and animals
Kinesin	Motor protein, moves toward the (+) end of MTs	Yeasts and animals

Centrosomes and SPBs

Sfi1	SPB protein, promotes SPB duplication in G1, until it targeted by Cdk to prevent SPB reduplication	Budding yeast
Spc42	SPB protein, targeted by Cdk and promotes SPB duplication at G1/S	Budding yeast

(Continued)

Table 9.1 (Continued)

Protein or complex	Function	System
Mps1	Mitotic kinase, needed for SPB duplication (yeast) and spindle checkpoint function	Yeasts and animals
Plk4	Polo-like kinase, necessary for centrosome duplication	Animals
Kinetochores		
CENP-A	Specialized centromeric histone H3 variant	Animals
Dam1	Outer kinetochore protein, contacts MTs	Budding yeast
Ndc80	Outer kinetochore protein	Budding yeast
Ska1	Outer kinetochore protein, contacts MTs	Animals
Aurora-B	Mitotic kinase, localized at the inner kinetochore	Animals
Polo kinase	Mitotic kinase	Animals
Survivin	Regulatory subunit of Aurora-B, targeted by M-Cdk	Animals
Spindle assembly checkpoint (SAC)		
Bub3	SAC component	Budding yeast
Mad2	SAC component	Budding yeast
Mad3	SAC component	Budding yeast
MCC	Bub3,Mad2,Mad3, Cdc20	Budding yeast

Table 9.1 (Continued)

Protein or complex	Function	System
Organelle inheritance		
Aurora-A	Mitotic kinase	Animals
GRASPs	Golgi ReAssembly and Stacking Proteins	Animals
Golgins	Golgi proteins important for stack assembly	Animals
GM130	Golgin, matrix protein targeted by M-Cdk	Animals
TOM	Mitochondrial import machinery	Budding yeast
Tom6	TOM component, targeted by Cdk	Budding yeast
Drp1	Mitochondrial fission regulator, targeted by M-Cdk	Animals
NPCs	Nuclear pore complexes	Yeasts and animals
Cytokinesis		
RhoA	GTPase, activator of the contractile ring assembly	Animals
GEFs	Guanine nucleotide-exchange factors	Yeasts and animals
Ect2	Guanine nucleotide-exchange factor involved in cytokinesis	Animals
Centralspindlin	Complex with motor properties, mediating the interaction of MTs and the Ect2 GEF in cytokinesis	Animals
Myosin-II	Nonmuscle myosin involved in cytokinesis	Animals
Actin	Building block of actin filaments	Yeasts and animals

(Continued)

Table 9.1 (Continued)

Protein or complex	Function	System
Other		
Tor	Kinase, master regulator of cellular anabolism and growth	Yeasts and animals
Protein phosphatase 1A	Phosphatase, with mitotic roles	Animals
Protein phosphatase 2A	Phosphatase, with mitotic roles	Animals
Cdc14	Phosphatase, activated at M/G1 and promoting mitotic exit	Budding yeast

References

1   Harris H. *The Birth of the Cell.* New Haven, Connecticut: Yale University Press; 1999 234 p.

2   Lagunoff D. A Polish, Jewish Scientist in 19th-Century Prussia. *Science.* 2002; 298(5602): 2331.

3   Flemming W. *Zellsubstanz, Kern und Zelltheilung.* Leipzig: Vogel; 1882.

4   Crow EW, Crow JF. 100 years ago: Walter Sutton and the chromosome theory of heredity. *Genetics.* 2002; 160(1): 1–4.

5   Stevens NM. *Studies in Spermatogenesis: With Especial Reference to the "Accessory Chromosome"*, vol. 36. Washington, DC: Carnegie Institution of Washington; 1905.

6   Sturtevant AH. *A History of Genetics.* Cold Spring Harbor, New York: Cold Spring Harbor Laboratory Press; 1965 174 p.

7   Boveri T. Concerning the origin of malignant tumours by Theodor Boveri. Translated and annotated by Henry Harris. *J Cell Sci.* 2008; 121(Supplement 1): 1–84.

8   Griffith F. The significance of pneumococcal types. *J Hyg (Lond).* 1928; 27(2): 113–59.

9   Avery OT, Macleod CM, McCarty M. Studies on the chemical nature of the substance inducing transformation of pneumococcal types: induction of transformation by a desoxyribonucleic acid fraction isolated from pneumococcus type III. *J Exp Med.* 1944; 79(2): 137–58.

10 Watson JD, Crick FH. Molecular structure of nucleic acids; a structure for deoxyribose nucleic acid. *Nature.* 1953; 171(4356): 737–8.

Two from One: A Short Introduction to Cell Division Mechanisms,
First Edition. Michael Polymenis.
© 2023 John Wiley & Sons Ltd. Published 2023 by John Wiley & Sons Ltd.

11 Howard A, Pelc S. Synthesis of deoxyribonucleic acid in normal and irradiated cells and its relation to chromosome breakage. *Heredity*. 1953; 6: 261–73.

12 Vatolina TY, Boldyreva LV, Demakova OV, Demakov SA, Kokoza EB, Semeshin VF, et al. Identical functional organization of nonpolytene and polytene chromosomes in *Drosophila melanogaster*. *PLoS One*. 2011; 6(10): e25960.

13 Neidhardt FC, Ingraham JL, Schaechter M. *Physiology of the Bacterial Cell*. Sunderland, MA: Sinauer Associates, Inc.; 1990.

14 Polymenis M, Kennedy BK. Unbalanced growth, senescence and aging. *Adv Exp Med Biol*. 2017; 1002: 189–208.

15 Bermudez RM, Wu PIF, Callerame D, Hammer S, Hu JC, Polymenis M. Phenotypic associations among cell cycle genes in *Saccharomyces cerevisiae*. *G3 Bethesda*. 2020; 10(7): 2345–51.

16 Godin M, Delgado FF, Son S, Grover WH, Bryan AK, Tzur A, et al. Using buoyant mass to measure the growth of single cells. *Nat Methods*. 2010; 7(5): 387–90.

17 Zangle TA, Teitell MA. Live-cell mass profiling: an emerging approach in quantitative biophysics. *Nat Methods*. 2014; 11(12): 1221–8.

18 Lange HC, Heijnen JJ. Statistical reconciliation of the elemental and molecular biomass composition of *Saccharomyces cerevisiae*. *Biotechnol Bioeng*. 2001; 75(3): 334–44.

19 Prescott DM. Relation between cell growth and cell division. III. Changes in nuclear volume and growth rate and prevention of cell division in *Amoeba proteus* resulting from cytoplasmic amputations. *Exp Cell Res*. 1956; 11(1): 94–8.

20 Pringle JR, Hartwell LH. The *Saccharomyces cerevisiae* cell cycle. In: *The Molecular and Cellular Biology of the Yeast Saccharomyces*. Cold Spring Harbor, New York: Cold Spring Harbor Laboratory Press; 1981. pp. 97–142.

21 Barbet NC, Schneider U, Helliwell SB, Stansfield I, Tuite MF, Hall MN. TOR controls translation initiation and early G1 progression in yeast. *Mol Biol Cell*. 1996; 7(1): 25–42.

22 Ewald JC, Kuehne A, Zamboni N, Skotheim JM. The yeast cyclin-dependent kinase routes carbon fluxes to fuel cell cycle progression. *Mol Cell*. 2016; 62(4): 532–45.

23 Kurat CF, Wolinski H, Petschnigg J, Kaluarachchi S, Andrews B, Natter K, et al. Cdk1/Cdc28-dependent activation of the major triacylglycerol lipase Tgl4 in yeast links lipolysis to cell-cycle progression. *Mol Cell.* 2009; 33(1): 53–63.

24 Prescott DM. Relations between cell growth and cell division. I. Reduced weight, cell volume, protein content, and nuclear volume of *Amoeba proteus* from division to division. *Exp Cell Res.* 1955; 9(2): 328–37.

25 Prescott DM. Relation between cell growth and cell division. II. The effect of cell size on cell growth rate and generation time in *Amoeba proteus. Exp Cell Res.* 1956; 11(1): 86–94.

26 Mitchison JM. Cell growth and protein synthesis. In: *The Biology of the Cell Cycle.* Cambridge, UK: Cambridge University Press; 1971. pp. 128–58.

27 Bryan AK, Engler A, Gulati A, Manalis SR. Continuous and long-term volume measurements with a commercial Coulter counter. *PLoS One.* 2012; 7(1): e29866.

28 Di Talia S, Skotheim JM, Bean JM, Siggia ED, Cross FR. The effects of molecular noise and size control on variability in the budding yeast cell cycle. *Nature.* 2007; 448(7156): 947–51.

29 Bryan AK, Goranov A, Amon A, Manalis SR. Measurement of mass, density, and volume during the cell cycle of yeast. *Proc Natl Acad Sci U S A.* 2010; 107(3): 999–1004.

30 Lee SS, Vizcarra IA, Huberts DHEW, Lee LP, Heinemann M. Whole lifespan microscopic observation of budding yeast aging through a microfluidic dissection platform. *Proc Natl Acad Sci.* 2012; 109(13): 4916–20.

31 Janssens GE, Meinema AC, Gonzalez J, Wolters JC, Schmidt A, Guryev V, et al. Protein biogenesis machinery is a driver of replicative aging in yeast. *elife.* 2015; 4: e08527.

32 Conlon I, Raff M. Differences in the way a mammalian cell and yeast cells coordinate cell growth and cell-cycle progression. *J Biol.* 2003; 2(1): 7.

33 Tzur A, Kafri R, LeBleu VS, Lahav G, Kirschner MW. Cell growth and size homeostasis in proliferating animal cells. *Science.* 2009; 325(5937): 167–71.

34 Son S, Tzur A, Weng Y, Jorgensen P, Kim J, Kirschner MW, et al. Direct observation of mammalian cell growth and size regulation. *Nat Methods.* 2012; 9(9): 910–2.

35   Son S, Kang JH, Oh S, Kirschner MW, Mitchison TJ, Manalis S. Resonant microchannel volume and mass measurements show that suspended cells swell during mitosis. *J Cell Biol.* 2015; 211(4): 757–63.

36   Vuaridel-Thurre G, Vuaridel AR, Dhar N, McKinney JD. Computational analysis of the mutual constraints between single-cell growth and division control models. *Adv Biosyst.* 2020; 4(2): e1900103.

37   Jun S, Taheri-Araghi S. Cell-size maintenance: universal strategy revealed. *Trends Microbiol.* 2015; 23(1): 4–6.

38   Soifer I, Robert L, Amir A. Single-cell analysis of growth in budding yeast and bacteria reveals a common size regulation strategy. *Curr Biol.* 2016; 26(3): 356–61.

39   Xie S, Skotheim JM. A G1 sizer coordinates growth and division in the mouse epidermis. *Curr Biol.* 2020; 30(5): 916–924.e2.

40   Sakaue-Sawano A, Yo M, Komatsu N, Hiratsuka T, Kogure T, Hoshida T, et al. Genetically encoded tools for optical dissection of the mammalian cell cycle. *Mol Cell.* 2017; 68(3): 626–640.e5.

41   Sakaue-Sawano A, Miyawaki A. Visualizing spatiotemporal dynamics of multicellular cell-cycle progressions with fucci technology. *Cold Spring Harb Protoc.* 2014; 2014(5).

42   Zielke N, Edgar BA. FUCCI sensors: powerful new tools for analysis of cell proliferation. *Wiley Interdiscip Rev Dev Biol.* 2015; 4(5): 469–87.

43   Bajar BT, Lam AJ, Badiee RK, Oh YH, Chu J, Zhou XX, et al. Fluorescent indicators for simultaneous reporting of all four cell cycle phases. *Nat Methods.* 2016; 13(12): 993–6.

44   Baserga R. *The Biology of Cell Reproduction.* Cambridge, MA: Harvard University Press; 1985.

45   Quastler H, Sherman FG. Cell population kinetics in the intestinal epithelium of the mouse. *Exp Cell Res.* 1959; 17(3): 420–38.

46   Hoffman JG. Theory of the mitotic index and its application to tissue growth measurement. *Bull Math Biophys.* 1949; 11(2): 139–44.

47   Mitchison JM. Single cell methods. In: *The Biology of the Cell Cycle.* Cambridge, UK: Cambridge University Press; 1971. pp. 18–9.

48 Blank HM, Callahan M, Pistikopoulos IPE, Polymenis AO, Polymenis M. Scaling of G1 duration with population doubling time by a cyclin in *Saccharomyces cerevisiae*. *Genetics*. 2018; 210(3): 895–906.

49 Pardee AB. G1 events and regulation of cell proliferation. *Science*. 1989; 246(4930): 603–8.

50 Guo J, Bryan BA, Polymenis M. Nutrient-specific effects in the coordination of cell growth with cell division in continuous cultures of *Saccharomyces cerevisiae*. *Arch Microbiol*. 2004; 182(4): 326–30.

51 Evans T, Rosenthal ET, Youngblom J, Distel D, Hunt T. Cyclin: a protein specified by maternal mRNA in sea urchin eggs that is destroyed at each cleavage division. *Cell*. 1983; 33(2): 389–96.

52 Hartwell LH, Culotti J, Pringle JR, Reid BJ. Genetic control of the cell division cycle in yeast. *Science*. 1974; 183(4120): 46–51.

53 Vassilev LT, Tovar C, Chen S, Knezevic D, Zhao X, Sun H, et al. Selective small-molecule inhibitor reveals critical mitotic functions of human CDK1. *Proc Natl Acad Sci U S A*. 2006; 103(28): 10660–5.

54 Blank HM, Perez R, He C, Maitra N, Metz R, Hill J, et al. Translational control of lipogenic enzymes in the cell cycle of synchronous, growing yeast cells. *EMBO J*. 2017; 36(4): 487–502.

55 Blank HM, Papoulas O, Maitra N, Garge R, Kennedy BK, Schilling B, et al. Abundances of transcripts, proteins, and metabolites in the cell cycle of budding yeast reveal coordinate control of lipid metabolism. *Mol Biol Cell*. 2020; 31(10): 1069–84.

56 Maitra N, He C, Blank HM, Tsuchiya M, Schilling B, Kaeberlein M, et al. Translational control of one-carbon metabolism underpins ribosomal protein phenotypes in cell division and longevity. *elife*. 2020; 9: e53127.

57 Nasmyth KA, Reed SI. Isolation of genes by complementation in yeast: molecular cloning of a cell-cycle gene. *Proc Natl Acad Sci U S A*. 1980; 77(4): 2119–23.

58 Reed SI, Hadwiger JA, Lörincz AT. Protein kinase activity associated with the product of the yeast cell division cycle gene CDC28. *Proc Natl Acad Sci U S A*. 1985; 82(12): 4055–9.

59 Nurse P. Genetic control of cell size at cell division in yeast. *Nature*. 1975; 256(5518): 547–51.

60 Nurse P, Thuriaux P. Regulatory genes controlling mitosis in the fission yeast *Schizosaccharomyces pombe*. *Genetics*. 1980; 96(3): 627–37.

61 Murray A. Paul Nurse and Pierre Thuriaux on wee mutants and cell cycle control. *Genetics*. 2016; 204(4): 1325–6.

62 Beach D, Durkacz B, Nurse P. Functionally homologous cell cycle control genes in budding and fission yeast. *Nature*. 1982; 300(5894): 706–9.

63 Lee MG, Nurse P. Complementation used to clone a human homologue of the fission yeast cell cycle control gene cdc2. *Nature*. 1987; 327(6117): 31–5.

64 Lohka MJ, Masui Y. Formation in vitro of sperm pronuclei and mitotic chromosomes induced by amphibian ooplasmic components. *Science*. 1983; 220(4598): 719–21.

65 Johnson RT, Rao PN. Mammalian cell fusion: induction of premature chromosome condensation in interphase nuclei. *Nature*. 1970; 226(5247): 717–22.

66 Masui Y, Markert CL. Cytoplasmic control of nuclear behavior during meiotic maturation of frog oocytes. *J Exp Zool*. 1971; 177(2): 129–45.

67 Wasserman WJ, Smith LD. The cyclic behavior of a cytoplasmic factor controlling nuclear membrane breakdown. *J Cell Biol*. 1978; 78(1): R15–22.

68 Gerhart J, Wu M, Kirschner M. Cell cycle dynamics of an M-phase-specific cytoplasmic factor in *Xenopus laevis* oocytes and eggs. *J Cell Biol*. 1984; 98(4): 1247–55.

69 Swenson KI, Farrell KM, Ruderman JV. The clam embryo protein cyclin A induces entry into M phase and the resumption of meiosis in *Xenopus* oocytes. *Cell*. 1986; 47(6): 861–70.

70 Minshull J, Blow JJ, Hunt T. Translation of cyclin mRNA is necessary for extracts of activated xenopus eggs to enter mitosis. *Cell*. 1989; 56(6): 947–56.

71 Murray AW, Kirschner MW. Cyclin synthesis drives the early embryonic cell cycle. *Nature*. 1989; 339(6222): 275–80.

72 Murray AW, Solomon MJ, Kirschner MW. The role of cyclin synthesis and degradation in the control of maturation promoting factor activity. *Nature*. 1989; 339(6222): 280–6.

73 Lohka MJ, Hayes MK, Maller JL. Purification of maturation-promoting factor, an intracellular regulator of early mitotic events. *Proc Natl Acad Sci U S A*. 1988; 85(9): 3009–13.

74 Gautier J, Norbury C, Lohka M, Nurse P, Maller J. Purified maturation-promoting factor contains the product of a *Xenopus* homolog of the fission yeast cell cycle control gene cdc2+. *Cell*. 1988; 54(3): 433–9.

75 Gautier J, Minshull J, Lohka M, Glotzer M, Hunt T, Maller JL. Cyclin is a component of maturation-promoting factor from *Xenopus*. *Cell*. 1990; 60(3): 487–94.

76 Pavletich NP. Mechanisms of cyclin-dependent kinase regulation: structures of Cdks, their cyclin activators, and Cip and INK4 inhibitors. *J Mol Biol*. 1999; 287(5): 821–8.

77 Surana U, Robitsch H, Price C, Schuster T, Fitch I, Futcher AB, et al. The role of CDC28 and cyclins during mitosis in the budding yeast *S. cerevisiae*. *Cell*. 1991; 65(1): 145–61.

78 Carter BL, Sudbery PE. Small-sized mutants of *Saccharomyces cerevisiae*. *Genetics*. 1980; 96(3): 561–6.

79 Sudbery PE, Goodey AR, Carter BL. Genes which control cell proliferation in the yeast *Saccharomyces cerevisiae*. *Nature*. 1980; 288(5789): 401–4.

80 Cross FR. DAF1, a mutant gene affecting size control, pheromone arrest, and cell cycle kinetics of *Saccharomyces cerevisiae*. *Mol Cell Biol*. 1988; 8(11): 4675–84.

81 Richardson HE, Wittenberg C, Cross F, Reed SI. An essential G1 function for cyclin-like proteins in yeast. *Cell*. 1989; 59(6): 1127–33.

82 Lew DJ, Dulić V, Reed SI. Isolation of three novel human cyclins by rescue of G1 cyclin (Cln) function in yeast. *Cell*. 1991; 66(6): 1197–206.

83 Gould KL, Nurse P. Tyrosine phosphorylation of the fission yeast cdc2+ protein kinase regulates entry into mitosis. *Nature*. 1989; 342(6245): 39–45.

84 Russell P, Nurse P. Negative regulation of mitosis by wee1+, a gene encoding a protein kinase homolog. *Cell*. 1987; 49(4): 559–67.

85 Spencer SL, Cappell SD, Tsai FC, Overton KW, Wang CL, Meyer T. The proliferation-quiescence decision is controlled by a bifurcation in CDK2 activity at mitotic exit. *Cell*. 2013; 155(2): 369–83.

86 Van TNN, Pellerano M, Lykaso S, Morris MC. Fluorescent protein biosensor for probing CDK/cyclin activity in vitro and in living cells. *ChemBioChem*. 2014; 15(15): 2298–305.

87 Ho B, Baryshnikova A, Brown GW. Unification of protein abundance datasets yields a quantitative *Saccharomyces cerevisiae* proteome. *Cell Syst*. 2018; 6(2): 192–205.e3.

88 Jumper J, Evans R, Pritzel A, Green T, Figurnov M, Ronneberger O, et al. Highly accurate protein structure prediction with AlphaFold. *Nature*. 2021; 596(7873): 583–9.

89 Varadi M, Anyango S, Deshpande M, Nair S, Natassia C, Yordanova G, et al. AlphaFold Protein Structure Database: massively expanding the structural coverage of protein-sequence space with high-accuracy models. *Nucleic Acids Res*. 2022; 50(D1): D439–44.

90 Humphreys IR, Pei J, Baek M, Krishnakumar A, Anishchenko I, Ovchinnikov S, et al. Computed structures of core eukaryotic protein complexes. *Science*. 2021; 374(6573): eabm4805.

91 Kõivomägi M, Valk E, Venta R, Iofik A, Lepiku M, Morgan DO, et al. Dynamics of Cdk1 substrate specificity during the cell cycle. *Mol Cell*. 2011; 42(5): 610–23.

92 Örd M, Loog M. How the cell cycle clock ticks. *Mol Biol Cell*. 2019; 30(2): 169–72.

93 Mendenhall MD, Jones CA, Reed SI. Dual regulation of the yeast CDC28-p40 protein kinase complex: cell cycle, pheromone, and nutrient limitation effects. *Cell*. 1987; 50(6): 927–35.

94 Tyers M. The cyclin-dependent kinase inhibitor p40SIC1 imposes the requirement for Cln G1 cyclin function at Start. *Proc Natl Acad Sci U S A*. 1996; 93(15): 7772–6.

95 de Nooij JC, Letendre MA, Hariharan IK. A cyclin-dependent kinase inhibitor, Dacapo, is necessary for timely exit from the cell cycle during *Drosophila* embryogenesis. *Cell*. 1996; 87(7): 1237–47.

96 Hadwiger JA, Wittenberg C, Mendenhall MD, Reed SI. The Saccharomyces cerevisiae CKS1 gene, a homolog of the Schizosaccharomyces pombe suc1+ gene, encodes a subunit of the Cdc28 protein kinase complex. *Mol Cell Biol*. 1989; 9(5): 2034–41.

97 Shah K, Liu Y, Deirmengian C, Shokat KM. Engineering unnatural nucleotide specificity for Rous sarcoma virus tyrosine kinase to uniquely label its direct substrates. *Proc Natl Acad Sci.* 1997; 94(8): 3565–70.

98 Hertz NT, Wang BT, Allen JJ, Zhang C, Dar AC, Burlingame AL, et al. Chemical genetic approach for kinase-substrate mapping by covalent capture of thiophosphopeptides and analysis by mass spectrometry. *Curr Protoc Chem Biol.* 2010; 2(1): 15–36.

99 Holt LJ, Tuch BB, Villén J, Johnson AD, Gygi SP, Morgan DO. Global analysis of Cdk1 substrate phosphorylation sites provides insights into evolution. *Science.* 2009; 325(5948): 1682–6.

100 Loog M, Morgan DO. Cyclin specificity in the phosphorylation of cyclin-dependent kinase substrates. *Nature.* 2005; 434(7029): 104–8.

101 Ubersax JA, Woodbury EL, Quang PN, Paraz M, Blethrow JD, Shah K, et al. Targets of the cyclin-dependent kinase Cdk1. *Nature.* 2003; 425(6960): 859–64.

102 Coudreuse D, Nurse P. Driving the cell cycle with a minimal CDK control network. *Nature.* 2010; 468(7327): 1074–9.

103 Swaffer MP, Jones AW, Flynn HR, Snijders AP, Nurse P. CDK substrate phosphorylation and ordering the cell cycle. *Cell.* 2016; 167(7): 1750–1761.e16.

104 Pirincci Ercan D, Chrétien F, Chakravarty P, Flynn HR, Snijders AP, Uhlmann F. Budding yeast relies on G1 cyclin specificity to couple cell cycle progression with morphogenetic development. *Sci Adv.* 2021; 7(23).

105 Basu S, Roberts EL, Jones AW, Swaffer MP, Snijders AP, Nurse P. The hydrophobic patch directs cyclin B to centrosomes to promote global CDK phosphorylation at mitosis. *Curr Biol.* 2020; 30(5): 883–892.e4.

106 Morgan DO. *The Cell Cycle: Principles of Control.* London, UK: New Science Press, Ltd; 2007.

107 Tyson JJ, Chen KC, Novak B. Sniffers, buzzers, toggles and blinkers: dynamics of regulatory and signaling pathways in the cell. *Curr Opin Cell Biol.* 2003; 15(2): 221–31.

108 Ingolia NT, Brar GA, Rouskin S, McGeachy AM, Weissman JS. The ribosome profiling strategy for

monitoring translation in vivo by deep sequencing of ribosome-protected mRNA fragments. *Nat Protoc.* 2012; 7(8): 1534–50.

109 Pomerening JR, Sontag ED, Ferrell JE. Building a cell cycle oscillator: hysteresis and bistability in the activation of Cdc2. *Nat Cell Biol.* 2003; 5(4): 346–51.

110 Sha W, Moore J, Chen K, Lassaletta AD, Yi CS, Tyson JJ, et al. Hysteresis drives cell-cycle transitions in *Xenopus laevis* egg extracts. *Proc Natl Acad Sci U S A.* 2003; 100(3): 975–80.

111 Morgan DO. Control of cell proliferation and growth. In: *The Cell Cycle: Principles of Control.* London, UK: New Science Press Ltd; 2007. p. 221.

112 Lindqvist A, van Zon W, Karlsson Rosenthal C, Wolthuis RMF. Cyclin B1-Cdk1 activation continues after centrosome separation to control mitotic progression. *PLoS Biol.* 2007; 5(5): e123.

113 Holt LJ, Krutchinsky AN, Morgan DO. Positive feedback sharpens the anaphase switch. *Nature.* 2008; 454(7202): 353–7.

114 López-Avilés S, Kapuy O, Novák B, Uhlmann F. Irreversibility of mitotic exit is the consequence of systems level feedback. *Nature.* 2009; 459(7246): 592–5.

115 Moseley JB, Mayeux A, Paoletti A, Nurse P. A spatial gradient coordinates cell size and mitotic entry in fission yeast. *Nature.* 2009; 459(7248): 857–60.

116 Keifenheim D, Sun XM, D'Souza E, Ohira MJ, Magner M, Mayhew MB, et al. Size-dependent expression of the mitotic activator Cdc25 suggests a mechanism of size control in fission yeast. *Curr Biol.* 2017; 27(10): 1491–1497.e4.

117 Connell-Crowley L, Elledge SJ, Harper JW. G1 cyclin-dependent kinases are sufficient to initiate DNA synthesis in quiescent human fibroblasts. *Curr Biol.* 1998; 8: 65–8.

118 Kõivomägi M, Swaffer MP, Turner JJ, Marinov G, Skotheim JM. G1 cyclin-Cdk promotes cell cycle entry through localized phosphorylation of RNA polymerase II. *Science.* 2021; 374(6565): 347–51.

119 Bertoli C, Skotheim JM, de Bruin RAM. Control of cell cycle transcription during G1 and S phases. *Nat Rev Mol Cell Biol.* 2013; 14(8): 518–28.

120 Bloom J, Cross FR. Multiple levels of cyclin specificity in cell-cycle control. *Nat Rev Mol Cell Biol*. 2007; 8(2): 149–60.

121 Litsios A, Huberts DHEW, Terpstra HM, Guerra P, Schmidt A, Buczak K, et al. Differential scaling between G1 protein production and cell size dynamics promotes commitment to the cell division cycle in budding yeast. *Nat Cell Biol*. 2019; 21(11): 1382–92.

122 Sommer RA, DeWitt JT, Tan R, Kellogg DR. Growth-dependent signals drive an increase in early G1 cyclin concentration to link cell cycle entry with cell growth. *elife*. 2021; 10: e64364.

123 Thorburn RR, Gonzalez C, Brar GA, Christen S, Carlile TM, Ingolia NT, et al. Aneuploid yeast strains exhibit defects in cell growth and passage through START. *Mol Biol Cell*. 2013; 24(9): 1274–89.

124 Zapata J, Dephoure N, Macdonough T, Yu Y, Parnell EJ, Mooring M, et al. PP2ARts1 is a master regulator of pathways that control cell size. *J Cell Biol*. 2014; 204(3): 359–76.

125 Polymenis M, Schmidt EV. Coupling of cell division to cell growth by translational control of the G1 cyclin CLN3 in yeast. *Genes Dev*. 1997; 11(19): 2522–31.

126 Valk E, Loog M. Multiple Pho85-dependent mechanisms control G1 cyclin abundance in response to nutrient stress. *Mol Cell Biol*. 2013; 33(7): 1270–2.

127 Schmoller KM, Turner JJ, Koivomagi M, Skotheim JM. Dilution of the cell cycle inhibitor Whi5 controls budding-yeast cell size. *Nature*. 2015; 526(7572): 268–72.

128 Zatulovskiy E, Zhang S, Berenson DF, Topacio BR, Skotheim JM. Cell growth dilutes the cell cycle inhibitor Rb to trigger cell division. *Science*. 2020; 369(6502): 466–71.

129 Dorsey S, Tollis S, Cheng J, Black L, Notley S, Tyers M, et al. G1/S transcription factor copy number is a growth-dependent determinant of cell cycle commitment in yeast. *Cell Syst*. 2018; 6(5): 539–554.e11.

130 Charvin G, Oikonomou C, Siggia ED, Cross FR. Origin of irreversibility of cell cycle start in budding yeast. *PLoS Biol*. 2010; 8(1): e1000284.

131 Skotheim JM, Di Talia S, Siggia ED, Cross FR. Positive feedback of G1 cyclins ensures coherent cell cycle entry. *Nature*. 2008; 454(7202): 291–6.

132 Ondracka A, Robbins JA, Cross FR. An APC/C-Cdh1 biosensor reveals the dynamics of cdh1 inactivation at the G1/S transition. *PLoS One*. 2016; 11(7): e0159166.

133 Sherr CJ, McCormick F. The RB and p53 pathways in cancer. *Cancer Cell*. 2002; 2(2): 103–12.

134 ICGC/TCGA Pan-Cancer Analysis of Whole Genomes Consortium. Pan-cancer analysis of whole genomes. *Nature*. 2020; 578(7793): 82–93.

135 Fassl A, Geng Y, Sicinski P. CDK4 and CDK6 kinases: from basic science to cancer therapy. *Science*. 2022; 375(6577): eabc1495.

136 Csardi G, Franks A, Choi DS, Airoldi EM, Drummond DA. Accounting for experimental noise reveals that mRNA levels, amplified by post-transcriptional processes, largely determine steady-state protein levels in yeast. *PLoS Genet*. 2015; 11(5): e1005206.

137 Spellman PT, Sherlock G, Zhang MQ, Iyer VR, Anders K, Eisen MB, et al. Comprehensive identification of cell cycle-regulated genes of the yeast *Saccharomyces cerevisiae* by microarray hybridization. *Mol Biol Cell*. 1998; 9(12): 3273–97.

138 Elliott SG, McLaughlin CS. Rate of macromolecular synthesis through the cell cycle of the yeast *Saccharomyces cerevisiae*. *Proc Natl Acad Sci U S A*. 1978; 75(9): 4384–8.

139 Tanenbaum ME, Stern-Ginossar N, Weissman JS, Vale RD. Regulation of mRNA translation during mitosis. *elife*. 2015; 4: e07957.

140 Yeeles JTP, Deegan TD, Janska A, Early A, Diffley JFX. Regulated eukaryotic DNA replication origin firing with purified proteins. *Nature*. 2015; 519(7544): 431–5.

141 Rhind N. DNA replication timing: random thoughts about origin firing. *Nat Cell Biol*. 2006; 8(12): 1313–6.

142 Diffley JF. On the road to replication. *EMBO Mol Med*. 2016; 8(2): 77.

143 Dahmann C, Diffley JF, Nasmyth KA. S-phase-promoting cyclin-dependent kinases prevent re-replication by inhibiting the transition of replication origins to a pre-replicative state. *Curr Biol*. 1995; 5(11): 1257–69.

144 Deegan TD, Diffley JF. MCM: one ring to rule them all. *Curr Opin Struct Biol*. 2016; 37: 145–51.

145 Gupta S, Friedman LJ, Gelles J, Bell SP. A helicase-tethered ORC flip enables bidirectional helicase loading. *elife*. 2021; 10: e74282.

146 Eaton ML, Galani K, Kang S, Bell SP, MacAlpine DM. Conserved nucleosome positioning defines replication origins. *Genes Dev*. 2010; 24(8): 748–53.

147 Frigola J, He J, Kinkelin K, Pye VE, Renault L, Douglas ME, et al. Cdt1 stabilizes an open MCM ring for helicase loading. *Nat Commun*. 2017; 8: 15720.

148 McGarry TJ, Kirschner MW. Geminin, an inhibitor of DNA replication, is degraded during mitosis. *Cell*. 1998; 93(6): 1043–53.

149 Wohlschlegel JA, Dwyer BT, Dhar SK, Cvetic C, Walter JC, Dutta A. Inhibition of eukaryotic DNA replication by geminin binding to Cdt1. *Science*. 2000; 290(5500): 2309–12.

150 Örd M, Venta R, Möll K, Valk E, Loog M. Cyclin-specific docking mechanisms reveal the complexity of M-CDK function in the cell cycle. *Mol Cell*. 2019; 75(1): 76–89.e3.

151 Mimura S, Seki T, Tanaka S, Diffley JFX. Phosphorylation-dependent binding of mitotic cyclins to Cdc6 contributes to DNA replication control. *Nature*. 2004; 431(7012): 1118–23.

152 Tanaka S, Umemori T, Hirai K, Muramatsu S, Kamimura Y, Araki H. CDK-dependent phosphorylation of Sld2 and Sld3 initiates DNA replication in budding yeast. *Nature*. 2007; 445(7125): 328–32.

153 Zegerman P, Diffley JFX. Phosphorylation of Sld2 and Sld3 by cyclin-dependent kinases promotes DNA replication in budding yeast. *Nature*. 2007; 445(7125): 281–5.

154 Douglas ME, Ali FA, Costa A, Diffley JFX. The mechanism of eukaryotic CMG helicase activation. *Nature*. 2018; 555(7695): 265–8.

155 Yeeles JTP, Janska A, Early A, Diffley JFX. How the eukaryotic replisome achieves rapid and efficient DNA replication. *Mol Cell*. 2017; 65(1): 105–16.

156 Aparicio JG, Viggiani CJ, Gibson DG, Aparicio OM. The Rpd3-Sin3 histone deacetylase regulates replication timing and enables intra-S origin control in *Saccharomyces cerevisiae*. *Mol Cell Biol*. 2004; 24(11): 4769–80.

157 Vogelauer M, Rubbi L, Lucas I, Brewer BJ, Grunstein M. Histone acetylation regulates the time of replication origin firing. *Mol Cell*. 2002; 10(5): 1223–33.

158 Kurat CF, Yeeles JTP, Patel H, Early A, Diffley JFX. Chromatin controls DNA replication origin selection, lagging-strand synthesis, and replication fork rates. *Mol Cell*. 2017; 65(1): 117–30.

159 Zhao J, Dynlacht B, Imai T, Hori T, Harlow E. Expression of NPAT, a novel substrate of cyclin E-CDK2, promotes S-phase entry. *Genes Dev*. 1998; 12(4): 456–61.

160 Zhao J, Kennedy BK, Lawrence BD, Barbie DA, Matera AG, Fletcher JA, et al. NPAT links cyclin E-Cdk2 to the regulation of replication-dependent histone gene transcription. *Genes Dev*. 2000; 14(18): 2283–97.

161 Guacci V, Koshland D, Strunnikov A. A direct link between sister chromatid cohesion and chromosome condensation revealed through the analysis of MCD1 in *S. cerevisiae*. *Cell*. 1997; 91(1): 47–57.

162 Michaelis C, Ciosk R, Nasmyth K. Cohesins: chromosomal proteins that prevent premature separation of sister chromatids. *Cell*. 1997; 91(1): 35–45.

163 Rao SSP, Huang SC, Glenn St Hilaire B, Engreitz JM, Perez EM, Kieffer-Kwon KR, et al. Cohesin loss eliminates all loop domains. *Cell*. 2017; 171(2): 305–320.e24.

164 Uhlmann F, Nasmyth K. Cohesion between sister chromatids must be established during DNA replication. *Curr Biol*. 1998; 8(20): 1095–101.

165 Haering CH, Farcas AM, Arumugam P, Metson J, Nasmyth K. The cohesin ring concatenates sister DNA molecules. *Nature*. 2008; 454(7202): 297–301.

166 Srinivasan M, Fumasoni M, Petela NJ, Murray A, Nasmyth KA. Cohesion is established during DNA replication utilising chromosome associated cohesin rings as well as those loaded de novo onto nascent DNAs. *elife*. 2020; 9: e56611.

167 Rolef Ben-Shahar T, Heeger S, Lehane C, East P, Flynn H, Skehel M, et al. Eco1-dependent cohesin acetylation during establishment of sister chromatid cohesion. *Science*. 2008; 321(5888): 563–6.

168 Unal E, Heidinger-Pauli JM, Kim W, Guacci V, Onn I, Gygi SP, et al. A molecular determinant for the establishment of sister chromatid cohesion. *Science*. 2008; 321(5888): 566–9.

169 Rankin S, Ayad NG, Kirschner MW. Sororin, a substrate of the anaphase-promoting complex, is required for sister chromatid cohesion in vertebrates. *Mol Cell.* 2005; 18(2): 185–200.

170 Hartwell LH, Weinert TA. Checkpoints: controls that ensure the order of cell cycle events. *Science.* 1989; 246(4930): 629–34.

171 Weinert TA, Hartwell LH. The RAD9 gene controls the cell cycle response to DNA damage in *Saccharomyces cerevisiae*. *Science.* 1988; 241(4863): 317–22.

172 Nasmyth K. Viewpoint: putting the cell cycle in order. *Science.* 1996; 274(5293): 1643–5.

173 Walker GC. Mutagenesis and inducible responses to deoxyribonucleic acid damage in *Escherichia coli*. *Microbiol Rev.* 1984; 48(1): 60–93.

174 Zhang Y, Hunter T. Roles of Chk1 in cell biology and cancer therapy. *Int J Cancer.* 2014; 134(5): 1013–23.

175 Zegerman P, Diffley JFX. Checkpoint-dependent inhibition of DNA replication initiation by Sld3 and Dbf4 phosphorylation. *Nature.* 2010; 467(7314): 474–8.

176 McIntosh JR, Hays T. A brief history of research on mitotic mechanisms. *Biology.* 2016; 5(4): 55.

177 Zernike F. How I discovered phase contrast. *Science.* 1955; 121(3141): 345–9.

178 Taylor EW. The mechanism of colchicine inhibition of mitosis. I. Kinetics of inhibition and the binding of H3-colchicine. *J Cell Biol.* 1965; 25(Suppl): 145–60.

179 Wells WA. The discovery of tubulin. *J Cell Biol.* 2005; 169(4): 552.

180 Weisenberg RC. Microtubule formation in vitro in solutions containing low calcium concentrations. *Science.* 1972; 177(4054): 1104–5.

181 Manka SW, Moores CA. Microtubule structure by cryo-EM: snapshots of dynamic instability. *Essays Biochem.* 2018; 62(6): 737–51.

182 Zhai Y, Borisy GG. Quantitative determination of the proportion of microtubule polymer present during the mitosis-interphase transition. *J Cell Sci.* 1994; 107(Pt 4): 881–90.

183 Karp G, Iwasa J, Marshall W. *Karp's Cell and Molecular Biology*. John Wiley & Sons; 2020 946 p.

184 Mitchison TJ. Polewards microtubule flux in the mitotic spindle: evidence from photoactivation of fluorescence. *J Cell Biol*. 1989; 109(2): 637–52.

185 Rieckhoff EM, Berndt F, Elsner M, Golfier S, Decker F, Ishihara K, et al. Spindle scaling is governed by cell boundary regulation of microtubule nucleation. *Curr Biol*. 2020; 30(24): 4973–4983.e10.

186 Roque H, Saurya S, Pratt MB, Johnson E, Raff JW. Drosophila PLP assembles pericentriolar clouds that promote centriole stability, cohesion and MT nucleation. *PLoS Genet*. 2018; 14(2): e1007198.

187 Scheer U. Historical roots of centrosome research: discovery of Boveri's microscope slides in Würzburg. *Philos Trans R Soc B Biol Sci*. 2014; 369(1650): 20130469.

188 Avena JS, Burns S, Yu Z, Ebmeier CC, Old WM, Jaspersen SL, et al. Licensing of yeast centrosome duplication requires phosphoregulation of Sfi1. *PLoS Genet*. 2014; 10(10): e1004666.

189 Elserafy M, Saric M, Neuner A, Lin TC, Zhang W, Seybold C, et al. Molecular mechanisms that restrict yeast centrosome duplication to one event per cell cycle. *Curr Biol*. 2014; 24(13): 1456–66.

190 Byers B, Goetsch L. Duplication of spindle plaques and integration of the yeast cell cycle. *Cold Spring Harb Symp Quant Biol*. 1974; 38: 123–31.

191 Keck JM, Jones MH, Wong CCL, Binkley J, Chen D, Jaspersen SL, et al. A cell cycle phosphoproteome of the yeast centrosome. *Science*. 2011; 332(6037): 1557–61.

192 Bettencourt-Dias M, Rodrigues-Martins A, Carpenter L, Riparbelli M, Lehmann L, Gatt MK, et al. SAK/PLK4 is required for centriole duplication and flagella development. *Curr Biol*. 2005; 15(24): 2199–207.

193 Aydogan MG, Wainman A, Saurya S, Steinacker TL, Caballe A, Novak ZA, et al. A homeostatic clock sets daughter centriole size in flies. *J Cell Biol*. 2018; 217(4): 1233–48.

194 Aydogan MG, Steinacker TL, Mofatteh M, Wilmott ZM, Zhou FY, Gartenmann L, et al. An autonomous oscillation

times and executes centriole biogenesis. *Cell.* 2020; 181(7): 1566–1581.e27.

195 Orlando DA, Lin CY, Bernard A, Wang JY, Socolar JES, Iversen ES, et al. Global control of cell-cycle transcription by coupled CDK and network oscillators. *Nature.* 2008; 453(7197): 944–7.

196 Rahi SJ, Pecani K, Ondracka A, Oikonomou C, Cross FR. The CDK-APC/C oscillator predominantly entrains periodic cell-cycle transcription. *Cell.* 2016; 165(2): 475–87.

197 Maiato H, Rieder CL, Khodjakov A. Kinetochore-driven formation of kinetochore fibers contributes to spindle assembly during animal mitosis. *J Cell Biol.* 2004; 167(5): 831–40.

198 Heald R, Tournebize R, Blank T, Sandaltzopoulos R, Becker P, Hyman A, et al. Self-organization of microtubules into bipolar spindles around artificial chromosomes in *Xenopus* egg extracts. *Nature.* 1996; 382(6590): 420–5.

199 Verma V, Maresca TJ. A celebration of the 25th anniversary of chromatin-mediated spindle assembly. *Mol Biol Cell.* 2022; 33(2): rt1.

200 Soares-de-Oliveira J, Maiato H. Mitosis: kinetochores determined against random search-and-capture. *Curr Biol.* 2022; 32(5): R231–4.

201 Renda F, Miles C, Tikhonenko I, Fisher R, Carlini L, Kapoor TM, et al. Non-centrosomal microtubules at kinetochores promote rapid chromosome biorientation during mitosis in human cells. *Curr Biol.* 2022; 32(5): 1049–1063.e4.

202 Monda JK, Cheeseman IM. The kinetochore-microtubule interface at a glance. *J Cell Sci.* 2018; 131(16): jcs214577.

203 Chen GY, Renda F, Zhang H, Gokden A, Wu DZ, Chenoweth DM, et al. Tension promotes kinetochore-microtubule release by Aurora B kinase. *J Cell Biol.* 2021; 220(6): e202007030.

204 Liu D, Vader G, Vromans MJM, Lampson MA, Lens SMA. Sensing chromosome bi-orientation by spatial separation of aurora B kinase from kinetochore substrates. *Science.* 2009; 323(5919): 1350–3.

205 Tsukahara T, Tanno Y, Watanabe Y. Phosphorylation of the CPC by Cdk1 promotes chromosome bi-orientation. *Nature.* 2010; 467(7316): 719–23.

206 Rieder CL, Cole RW, Khodjakov A, Sluder G. The check-point delaying anaphase in response to chromosome monoorientation is mediated by an inhibitory signal produced by unattached kinetochores. *J Cell Biol.* 1995; 130(4): 941–8.

207 Musacchio A. The molecular biology of spindle assembly checkpoint signaling dynamics. *Curr Biol.* 2015; 25(20): R1002–18.

208 Chao WCH, Kulkarni K, Zhang Z, Kong EH, Barford D. Structure of the mitotic checkpoint complex. *Nature.* 2012; 484(7393): 208–13.

209 Blank HM, Maitra N, Polymenis M. Lipid biosynthesis: when the cell cycle meets protein synthesis? *Cell Cycle.* 2017; 16(10): 905–6.

210 Meseroll RA, Cohen-Fix O. The malleable nature of the budding yeast nuclear envelope: flares, fusion, and fenestrations. *J Cell Physiol.* 2016; 231(11): 2353–60.

211 Maitra N, Hammer S, Kjerfve C, Bankaitis VA, Polymenis M. Translational control of lipogenesis links protein synthesis and phosphoinositide signaling with nuclear division in *Saccharomyces cerevisiae. Genetics.* 2022; 220(1): iyab171.

212 Chan YH, Marshall WF. Scaling properties of cell and organelle size. *Organogenesis.* 2010; 6(2): 88–96.

213 Carlton JG, Jones H, Eggert US. Membrane and organelle dynamics during cell division. *Nat Rev Mol Cell Biol.* 2020; 21(3): 151–66.

214 Glick BS. Can the Golgi form de novo? *Nat Rev Mol Cell Biol.* 2002; 3(8): 615–9.

215 Ayala I, Mascanzoni F, Colanzi A. The Golgi ribbon: mechanisms of maintenance and disassembly during the cell cycle. *Biochem Soc Trans.* 2020; 48(1): 245–56.

216 Lowe M, Rabouille C, Nakamura N, Watson R, Jackman M, Jämsä E, et al. Cdc2 kinase directly phosphorylates the cis-Golgi matrix protein GM130 and is required for Golgi fragmentation in mitosis. *Cell.* 1998; 94(6): 783–93.

217 Lill R, Freibert SA. Mechanisms of mitochondrial iron-sulfur protein biogenesis. *Annu Rev Biochem.* 2020; 89(1): 471–99.

218 Harbauer AB, Opalińska M, Gerbeth C, Herman JS, Rao S, Schönfisch B, et al. Mitochondria. Cell cycle-dependent

regulation of mitochondrial preprotein translocase. *Science.* 2014; 346(6213): 1109–13.

219 Shiota T, Traven A, Lithgow T. Mitochondrial biogenesis: cell-cycle-dependent investment in making mitochondria. *Curr Biol.* 2015; 25(2): R78–80.

220 Rafelski SM, Viana MP, Zhang Y, Chan YHM, Thorn KS, Yam P, et al. Mitochondrial network size scaling in budding yeast. *Science.* 2012; 338(6108): 822–4.

221 Mishra P, Chan DC. Mitochondrial dynamics and inheritance during cell division, development and disease. *Nat Rev Mol Cell Biol.* 2014; 15(10): 634–46.

222 Moore AS, Coscia SM, Simpson CL, Ortega FE, Wait EC, Heddleston JM, et al. Actin cables and comet tails organize mitochondrial networks in mitosis. *Nature.* 2021; 591(7851): 659–64.

223 Chan YHM, Reyes L, Sohail SM, Tran NK, Marshall WF. Organelle size scaling of the budding yeast vacuole by relative growth and inheritance. *Curr Biol.* 2016; 26(9): 1221–8.

224 Yabe I, Horiuchi K, Nakahara K, Hiyama T, Yamanaka T, Wang PC, et al. Patch clamp studies on V-type Atpase of vacuolar membrane of haploid *Saccharomyces cerevisiae*: preparation and utilization of a giant cell containing a giant vacuole. *J Biol Chem.* 1999; 274(49): 34903–10.

225 Han BK, Aramayo R, Polymenis M. The G1 cyclin Cln3p controls vacuolar biogenesis in *Saccharomyces cerevisiae. Genetics.* 2003; 165(2): 467–76.

226 Han BK, Bogomolnaya LM, Totten JM, Blank HM, Dangott LJ, Polymenis M. Bem1p, a scaffold signaling protein, mediates cyclin-dependent control of vacuolar homeostasis in *Saccharomyces cerevisiae. Genes Dev.* 2005; 19(21): 2606–18.

227 Jin Y, Weisman LS. The vacuole/lysosome is required for cell-cycle progression. *elife.* 2015; 4: e08160.

228 Dey G, Culley S, Curran S, Schmidt U, Henriques R, Kukulski W, et al. Closed mitosis requires local disassembly of the nuclear envelope. *Nature.* 2020; 585(7823): 119–23.

229 Pollard TD, O'Shaughnessy B. Molecular mechanism of cytokinesis. *Annu Rev Biochem.* 2019; 88: 661–89.

230 Garno C, Irons ZH, Gamache CM, McKim Q, Reyes G, Wu X, et al. Building the cytokinetic contractile ring in an early embryo: Initiation as clusters of myosin II, anillin and septin, and visualization of a septin filament network. *PLoS One*. 2021; 16(12): e0252845.

231 Shuster CB, Burgess DR. Parameters that specify the timing of cytokinesis. *J Cell Biol*. 1999; 146(5): 981–92.

232 Cao LG, Wang YL. Signals from the spindle midzone are required for the stimulation of cytokinesis in cultured epithelial cells. *Mol Biol Cell*. 1996; 7(2): 225–32.

233 Wagner E, Glotzer M. Local RhoA activation induces cytokinetic furrows independent of spindle position and cell cycle stage. *J Cell Biol*. 2016; 213(6): 641–9.

234 Niiya F, Xie X, Lee KS, Inoue H, Miki T. Inhibition of cyclin-dependent kinase 1 induces cytokinesis without chromosome segregation in an ECT2 and MgcRacGAP-dependent manner. *J Biol Chem*. 2005; 280(43): 36502–9.

235 Miyazaki M, Chiba M, Eguchi H, Ohki T, Ishiwata S. Cell-sized spherical confinement induces the spontaneous formation of contractile actomyosin rings in vitro. *Nat Cell Biol*. 2015; 17(4): 480–9.

236 Rappaport R. Cell division: direct measurement of maximum tension exerted by furrow of echinoderm eggs. *Science*. 1967; 156(3779): 1241–3.

237 Kukhtevich IV, Lohrberg N, Padovani F, Schneider R, Schmoller KM. Cell size sets the diameter of the budding yeast contractile ring. *Nat Commun*. 2020; 11(1): 2952.

238 Pollard TD. Nine unanswered questions about cytokinesis. *J Cell Biol*. 2017; 216(10): 3007–16.

239 Atilla-Gokcumen GE, Muro E, Relat-Goberna J, Sasse S, Bedigian A, Coughlin ML, et al. Dividing cells regulate their lipid composition and localization. *Cell*. 2014; 156(3): 428–39.

Index

Page locators in **bold** indicate tables. Page locators in *italics* indicate figures. This index uses letter-by-letter alphabetization.

Two from One: A Short Introduction to Cell Division Mechanisms, First Edition. Michael Polymenis.
© 2023 John Wiley & Sons Ltd. Published 2023 by John Wiley & Sons Ltd.

Printed in the USA
CPSIA information can be obtained
at www.ICGtesting.com
CBHW080011050624
9578CB00022B/300